Don Coyote

Don Coyote

The Good Times and the Bad Times of a Much Maligned American Original

DAYTON O. HYDE

Johnson Books
BOULDER

Published by Johnson Books, a division of Johnson Publishing
Company, 1880 South 57th Court, Boulder, Colorado 80301.
Visit our website at www.JohnsonBooks.com.
E-mail: books@jpcolorado.com.

9   8   7   6   5   4   3   2   1

Cover design: Debra B. Topping
Cover photo: Stephen Collector

Library of Congress Cataloging-in-Publication Data
Hyde, Dayton O., 1925–
    Don coyote: the good times and the bad times of a much maligned American
original / Dayton O. Hyde.
        p. cm.
    ISBN 1-55566-355-9
    1. Coyote—Behavior. 2. Hyde, Dayton O., 1925– 3. Mammals—Behavior. 4.
Mammals—Behavior—Oregon. 5. Ranchers—Oregon—Biography. I. Title
    QL737.C22H93 2004
    599.77'25—dc22   2004010442

Printed in the United States by
Johnson Printing
1880 South 57th Court
Boulder, Colorado 80301

Printed on ECF paper with soy ink

1

I DON'T KNOW WHY that particular coyote bothered me so. For some reason it wouldn't let me alone. I was doing some tractor work on our Oregon cattle ranch—mending ditch banks, putting in culverts, cleaning up winter windfalls—and rattling along on an ancient wired and welded tracklayer of flaking rust and miserable parts that should have been keeping Lindbergh's *Spirit of St. Louis* company in the Smithsonian.

The Allis Chalmers shook, squealed, clanked, and wheezed so hard it hurt my teeth, rendered me deaf, jolted my spine, and made me a nervous wreck wondering which gasp would be its last and who would finish paying the mortgage on it after we both were gone.

Every morning the coyote would trot from the lodgepole thickets along the river and follow me across the meadows, not forty feet behind, either stupidly curious or just plain craving company with that snorting monster of a tractor which growled across its territory, sent ground squirrels tumbling into their holes, and polluted the mountain air with black clouds of diesel smoke.

That the coyote was around at all was probably because we post our ranch against trespassing. This part of southern Oregon, occupied by a dozen cattle and sheep to every human, had been subjected to every known form of predator control except consideration: traps, poisons, snowmobiles, rifles, clubs, helicopters, rabbit distress calls, smoke bombs, poison gas, twisted barbed wire, cyanide, guns, dynamite, voodoo. You name it, it had been tried.

In the rear windows of their pick-up trucks, most of my neighbors carry the local status symbol—a loaded rifle for coyotes, rustlers, and other varmints—while the usual conversation at the local barber shop centers on low prices for farm products, depredation by predators, and the failure of the "damn governmint" to do a "damn thing" about either.

Off my ranch, coyotes had been shot at so much that I saw them only as streaks of tattered brown, vamoosing into forest or sage, apparently sensing that for all his powers Man can't shoot through trees or hit something he can no longer see.

Either this one sensed my respect for wildlife or was just plain loco. I tried to ignore it, but after a week of its supervision and its persistent digging in my nice green meadow, I became increasingly neurotic, fancying that the animal mocked me when the tractor mired in mud, ran out of fuel, failed to start, or overheated in a hiccuping volcano of antifreeze-scented steam. And when the animal dug up a meadow mouse in a brand new dike and sent irrigation water coursing out over the meadow I'd just dried out for haying, I shook my fist at him.

"Get off my case or I'll bust a wheel on your wagon, Mister Smart!" I threatened.

Bang! Bang! The tractor popped as though firing off a

few rounds of its own, but the animal didn't seem to notice. Bang! The tractor stopped with such a lurch that I flattened my nose against the oil gauge. The coyote turned its face away as though hiding a smile. No amount of mechanical sorcery on my part could persuade Big Alice out of her balk, so I set off afoot for the ranch house, with the coyote watching me from atop a knoll. It cocked its head as it eyed first me and then the tractor. To the animal p\[e\]rhaps I was no more than a part of the machine that ha\[d\] jiggled loose and was rolling down the hill on its own.

The next \[d\]ay I drove the sixty miles to Klamath Falls for tractor \[pa\]rts. Just as I bustled past the local sporting goods stor\[e\] \[o\]ut came one of my neighbors packing a brand new \[3\]0 rifle with buckhorn sights, a stock sleek as a frog'\[s\] \[a\]nd a whole box of ammunition.

"Here,\[" ... sai\]d. "I bought this just for you. You can't \[... o\]f a pig, so on the way home demolish a few beer \[... s\]harpen up. The other day I was cuttin' across y\[... on\] horseback and seen a coyote needs a comeup\[... \] the gun and use it!"

He s\[...\] \[the\] rifle into my arms, setting me back on my he\[... an\]d before I could refuse it, he was halfway down \[... b\]lock.

A w\[oma\]n stared at me as though she had just sighted a presid\[ential\] assassin, and shrank back into a doorway.

"B\[... birthd\]ay present for my kid," I said, holding the rifle awk\[wardly ...\] so she'd see I didn't know how to use it. On the \[... bac\]k to my pick-up, I took the back alleys and hid \[... b\]ehind the seat.

I \[... sha\]ken by my neighbor's hostility. If I didn't sho\[... al\]l with his gift, he'd banish me forever as a f\[... b\]eing alone, but not that alone. It's nice to kn\[ow ... th\]ere beyond all those ridges there is civili\[zation ... some\]one to help you in a bind, someone who th\[... \] pretty regular guy. The man was angry

at my coyote, and I had to admit to myself, it had
crawled under my skin a little too.

All the way back to the ranch I gave myself attitude
talks, visualizing that smug monster of a coyote as it ambled out of my woods to follow me again. "In complete
control of the situation. Working its witchcraft . . . Out
for a stroll and a good laugh at Mankind, with nothing to
fear from this dumb rancher . . . Picking out the fattest
calves for itself as though it held the first mortgage on the
ranch . . . Biding his time to make a raid . . . Well, it
was in for a shock!"

"Going to get that coyote that's always following the
tractor," I told the family at breakfast.

My children—Dayton, Ginny, Marsha, John, and Taylor—glanced at me skeptically.

"If you're going to shoot a gun," Dayton said, "I'll go
stand next to the coyote. That might be the safest place
on the ranch."

"You act so mean, Dad," Marsha said. "You're really
just an old teddy bear, you know."

The rest smirked at each other and went on chewing
their sourdough hotcake cuds. Bunch of softies, those
kids. The thought of my neighbor's anger and memories
of all the favors he'd done me kept my resolve alive.

"Teach that coyote not to kill sheep," I said.

"I didn't know we ran any sheep," my wife, Gerdi,
muttered under her breath, deferring a little to the king,
but only as one gives in patiently to an idiot rather than
create a scene.

"Teach it not to kill the sheep it wishes we had," I said
stubbornly.

I rose abruptly from the table. Rather than eat with
such a table of coyote lovers, I stalked off down to the
barn, kicking pinecones as I went.

Once I gained the tractor, I laid the loaded carbine on
Big Alice's battered pads, bolted on new fuel injectors,
set the rockers, then went through my usual ritual—

checking the radiator for water, wiping the oil dipstick on my Levi pant leg, standing on my head trying to insert the stick back in the hole, and greasing such zerk fittings as had not jiggled off during the previous running.

The big coyote moseyed out from the edge of the timber, sat patiently, and yawned as though to show me I was late. There was a piece of sagebrush caught in its pelage along the ribs where it had been curled up waiting for me. Its eyes were amber, and its smile Ultra Brite white. I was that close. Wind played through the hairs of its coat.

I tried tightening some bolts on the tractor and kept losing my wrench down into the bowels of the machine. I thought of photographs I'd seen of dead sheep. Some of my sheepmen friends seemed constantly to carry such photographs in their pockets. I made up my mind that I should shoot the animal not for myself, but for my neighbors. Sheepmen had mortgages too.

I felt sick to my stomach. Slowly, cautiously, pretending to inspect the machine, I lifted the rifle from the steel tracks. Silently, I turreted the barrel about, then sighted in on the coyote's body, cocking the hammer with trembling thumb.

"You're just a big bluff! You wouldn't shoot; you know you wouldn't," some inner voice taunted me.

"Oh, wouldn't I?" I quarreled with ever stiffening resolve.

Uncocking, I laid the rifle back down on the track pads. I fancied driving the tractor even closer. No sense risking the faint chance I might miss His Arrogance. I'd never get such a golden opportunity again.

It lay down on the grass and rolled, four paws up in the air, wriggling happily as it scratched its back. It looked at me curiously, with its eyes upside down. "Woof!" it barked, as though startled by what it saw, then righted itself and lay panting, pink tongue drooping as though even that little bit of action had winded it.

I felt my anger melting but gritted my teeth. "Just you wait, Mister Sneak!" I threatened under my breath. "I'll hang your miserable carcass from a fence for the ravens to pick. But first I've got to drive closer."

Pretending to ignore it, but casting sidelong glances as though hoping to catch it in some flagrant act, I eased up into the tractor seat, cranked the engine into new life, waited calmly for the smoke to clear and the oil pressure to build, then pulled back the hand clutch.

Protesting, the big track plates clanked, shrieked, but yielded and followed each other obediently forward. All thirty-five tons of Allis Chalmers steel began to move.

"That's the way, Big Alice, baby," I murmured.

Above the roar of the laboring machine, I heard a sudden, unfamiliar clank and creak. Too late I realized I'd left my neighbor's carbine on the tracks. As the machine rumbled forward, it laid my friend's brand new purchase on the ground ahead and proceeded to run the length of it.

As I clambered down to inspect the ruin, the coyote did more than wag its tail: It tipped back its head, and in a startling chorus of yips, yowls, yaps, and yelps, it sat there laughing at me, I could swear it.

Relief came flooding over me. My children were right about me! Plunking myself down beside the remains of Winchester's best, I sat there laughing with the animal.

2

IT WAS A WEEK before I saw the animal again—seven hard, busy days spent on a section of the ranch known as Calimus, patching the outside fences to hold our yearling cattle for the summer. Hardly the best part of the ranch, it was a thousand-acre forest and meadow parcel acquired, without much thought given to it, when I purchased the main ranch, Yamsi, from my uncle. It was in another watershed, a dozen mountainous miles away from ranch headquarters, and had no ranch house, only an old cowboy line camp and a pump house, both long ago abandoned to pack rats and porcupines.

I had climbed a rimrock overlooking the lower part of the Calimus field when I saw a coyote mousing in a late snow. It was almost evening, and the color off toward Saddle Mountain was washed with wine. I had my camera with me and had used up nearly all my film on the animal when it threw back its head and started to howl.

There was only one frame left, but I captured it silhouetted against the vast loneliness of its range.

There was an air of mystery about that Calimus field. Ancient Klamath Indian legend held it as a place to be avoided, a valley where giant bears devoured the unwary. It was a place of banishment perhaps, for such stone-age implements as one finds are crude, as though the maker was only biding his time until death. In the lower end was a bottom of gloomy lodgepole pines so thick one could become lost in them. Even the deer seemed to keep to the ridges.

We summered our replacement heifers there, pumped water for them with a big Fairbanks Morse single-cylinder engine that shot blue donuts of exhaust smoke down the valley and drank a gallon of gasoline for every gallon of water it pumped. In the late thirties Ewauna Lumber Company built a lumber camp on a pumice hillside, but that community too is gone. Only some rusty cans, catsup bottles, tumbled privies, pitch-encrusted boards, and broken children's toys give testimony that folks once lived and loved there.

During the pasture season, the heifers got lost in the thickets and small grassy glades and pretty much fended for themselves. To find them you tracked them or caught them on water, glad to get them back safely in the fall. They were fat but as wild as though they had spent the summer being afraid. It was a place of goshawks and great gray owls, birds of my own solitary inclination.

Lonely though it was, I loved it, for there I could be alone as in no place else on earth. And someday it would play a major role in my life.

This particular year the fences were in bad shape, broken down by fallen trees, heavy snows, and running deer. I wanted to get back to my new friend, the coyote, but the work went slowly. Try to stretch a rusty length of barbed wire, and somewhere up the line it would break on you as though an Indian spirit had cut it.

Hard as the fencing was, it was a relief not to run the old tractor. I had deserted Big Alice where she had failed me most recently, on a timbered knoll overlooking the Williamson River, a gentle, clear stream gushing from a fault line along the base of Yamsay Mountain and coursing slowly north through the ranch in a series of horseshoe bends. Busy with Calimus fences, I had neither time nor patience for the old machine's tantrums.

One Sunday I took the day off to ride down through the ranch. Mounted on my favorite holiday treat, a mare named Straight Edge, who'd borne the outfit many a foal and who'd been with me half my life, I trotted past Big Alice. Out from under the silent tractor came the big coyote, and sitting on its haunches, it eyed me curiously.

It was obvious that by now it had taken full proprietary rights to the hulking wreck, for at my approach on the elegant old bay mare, it rose to its feet and, gray hackles bristling, trotted around the tractor in a circle, jacking up a hind leg at each corner of the tracks and hosing it down with a drench of urine.

Catching sight of the coyote, Straight Edge snapped out of her reverie with a snort of terror and shied so violently her front legs tangled like hop vines: then down we both went in a heap.

Such uncalled-for commotion startled the coyote. Losing its cool, it ran off through the woods, stopping to look back only to take off again. By way of apology, I left it a ham and mayonnaise sandwich—half my lunch—on a nearby rock, and when I rode past an hour later, the food was gone; a brand new diggery in my nice clover meadow hinted that the coyote had accepted my gift and buried it, perhaps to improve the taste.

I whistled, hoping it might step out from its tractor lair, but either it had made its human contact for the day and was sleeping off the experience or it was away doing whatever it is coyotes do for a living.

Now that the animal had won my friendship, the do-

nation of half my lunch became an almost daily ritual.
But this was such a blatant betrayal of local ranching sen-
timent, I first looked over my shoulder to see that none
of the neighboring livestock folk were riding by.

On the surface, such precaution may have seemed silly
on my part, but it was based on a realistic fear. I felt that
as soon as it was buzzed about the community that I had
sided with a coyote, I would lose friendships I really valued, for in this part of the country the coyote issue is as
emotional as that of gun control, and there are all too few
friendships that span both sides of the argument. By aid-
ing and abetting a coyote, I would be standing against
my own industry. Even though there were many
ranchers who privately agreed with my attitude toward
predators, the cattle and sheep associations had closed
ranks to stump for more predator control.

It was an alliance of the rural landowners against the
urban nonlanded majority, with the rural population
standing on principle against a powerful urban force that
had saddled them with one new tax, one new law, one
environmental restriction after another. There were not
that many rural votes left, and so, one country faction
sided with another.

Even though many cattlemen had little or no interest in
predator control, if the sheepmen wanted it and urban
forces were against it, then, by hell, the cattlemen would
back the rural alliance a hundred percent. The rural associations tended toward bloc thinking and bloc action. A
slight disagreement with the industry on philosophy
could brand one as an Audubon or, worse yet, an environmentalist.

The next time I met a sheepman on the streets of
Klamath Falls, I felt so uncomfortable that I averted my
glance. "He knows!" I thought. "He knows I've been
nice to a coyote!" By the time I'd walked a hundred
yards, the very realization that I was concerned over what
other people felt about me began to anger me. I was a

cattleman, and cattlemen were noted for their independence. Then why were we so dependent upon what other ranch people thought of us? I began to grow angry, first, at my friends, who wouldn't allow me an opinion of my own, and second, at the folks in cities, who, having formed their own bloc, had made a rural coalition necessary in the first place. By the time I had walked another hundred yards, I had worked myself into a real rage.

Hell, it was my land, not theirs! All six thousand acres of it! And the place was thirty miles from town. If I wanted coyotes on my property, I'd damn well have them.

Suddenly, there were my friends Joe and Lena driving toward me up Main Street. Owned lots of sheep as well as cattle. I couldn't think of a couple I respected more. They'd be hurt if I didn't wave. I compromised by raising my hand in what might have passed either for a wave or an ear scratch and kept on walking, acting in a hurry so they wouldn't find a parking space and demand that I join them for coffee.

By the time the next sheepmen came along, a couple of marvelous Basques with whom I'd hoisted a leather bottle of sun-heated wine many a time, I'd made a final decision and raised my tent in the coyote camp. Intent upon window displays, I sailed on past.

"Hey," one of them called. "Come have a glass of Red Mountain with us!"

"I'm late," I said. "I've got to run." I was turning them down before they had a chance to turn me down, trying to prepare myself for an ostracism I knew had to come. For years I had been active in state cattlemen's organizations, and I had a hunch the chairmanships of committees I headed would soon be given to someone more in tune with industry thinking. With me, however, there was no such thing as being a closet coyote lover. If I wasn't quite a natural loner, standing up for coyotes and allowing them a refuge on my ranch would now make me one. Perhaps I could have just left them there and been quiet about it, but I wasn't a man to be silent where matters of principle were involved.

More and more, in the ensuing weeks, I built my working days around the ritual of feeding that wild beast. Ranch cooks are cranky by profession; my wife was no exception. Had I asked for an extra sandwich, it would have been tantamount to suggesting that I hadn't been getting enough before. So slight a comment on the food had sent many a good cook packing for town, and I didn't want to chance losing mine.

And so, rather than ask for more, I gave the coyote half my lunch and went hungry. With my hard, active work schedule, I became more irritable than ever. But still my friend got its handout, whether or not it deigned to eat it in my company.

This time of year, I should have been riding all the fields, poking endlessly through the cattle, painting iodine on the wet, pink navels of newborn calves to guard against infection, or roping older calves to force a scour pill down a protesting throat. But because the coyote lived at the south end of the ranch, the north end became a no-man's-land where sick cattle fevered and got well on their own antibodies without much medicinal help from me. Irrigation canals silted in, made deltas, then created new channels, while the old ones grew grass or turned into deserts as they pleased.

The old mare came close to being able to read my thoughts. Without my touching a rein, she bore me out of the corrals of a morning and headed straight for the silent tractor where, before I could dismount and in fla-grant violation of all tenets of her cow horse training, she dropped her head to graze. According to ranch ethics, any time a western horse got its head that low while being ridden, it had better either be drinking from a stream or fixing to throw you off. But grazing? NEVER!

Often the coyote was at the tractor waiting for me.

Beneath Big Alice's hulk it had found himself an impregnable stronghold, protected from storms by tracks on two sides, a bulldozer blade in front, and the massive iron fortress overhead. In such a bastion, it could weather any attack, whether from pursuing dogs, hovering helicopters, or volleys of rifle fire. I could have used the tractor elsewhere, but somehow to repair the engine and move it off down the valley seemed to violate the new friendship I was trying so hard to cultivate. I remained steadfast even though I sweated while moving dirt with a shovel, each blister a medal of honor, or spent hours at some project the giant machine could have handled with one swipe of the blade.

It wasn't long before my family realized I was neglecting my work. From Taylor, the youngest, on up to Dayton, diverse as they might be in interests and personalities, one great trait their mother had passed on to them was that they couldn't stand to see the ranch less than perfectly managed. Having a father whose priorities might place watching a pair of sandhill cranes dancing on a meadow above fixing a broken fence was hard for them to accept.

"Relax," I told them when they got on my case. "You're only here this go-round. It's hard to see beauty when you're always on the run. Work will always be there tomorrow, but those sandhill cranes raising their young on the marshes, those Wilson's snipe winnowing over the river, that coyote hunting mice unafraid on the meadow, who knows how long you'll be able to enjoy sights like those?"

One night at supper my daughter Ginny regarded me over her soup plate. "What's wrong with the monster tractor this time, Dad?"

I almost choked on a piece of meat and took a long time to settle down. Truth was, there was an air bubble in the secondary fuel filter. Five minutes with a monkey wrench could have fixed it.

"Needs a new set of bearings in the dinkerator chuck,"

I explained to the whole family. "Parts-man told me he might have to send clear back East." I was dying to share my excitement that the coyote had a den beneath it, but the truth just wouldn't out.

It was Dayton's turn to single me out for attack. He was silent for a moment, as though brooding over how to put me up for adoption. "Better move that rig soon," he cautioned. "Saw a coyote today lifting his leg at all four corners trying to wash it out of his way. We may get even more rust on the machine than we already have. Think I'll collect that coyote's pelt one of these days for a collar on my winter coat. What did you do with the rifle you came home with? Lose it? Haven't seen it around."

His bemused stare held me in a grip I couldn't shake.

"It's around," I mumbled. "If there's a coyote on this place needs getting, I'll get it myself. Everybody else, hands off!"

I was being baited, and I knew it. Not one of my kids would harm an animal. They grinned into their bowls and let me be.

"What good does a coyote do?" Marsha asked the next morning at breakfast, as though the presence of a coyote on the ranch had to be justified in terms of efficiency.

"What good does a coyote do?" I stalled. "Well, for one thing, it kills sheep." Ever since the range wars when I was a boy, I'd never been too fond of the woolies anyway.

"No, Dad. Really and truly."

"The coyote," I said, "is the veterinarian of the sagebrush. It devours the sick, the lame, and the genetically unfit, keeping the rest of the population healthy. It also eats mice, ground squirrels, grasshoppers, jackrabbits, and carrion." Pretty fast thinking for a guy who only recently had become a coyote convert.

"But how many?" Marsha persisted. "How does it all add up? Do they do as much good as harm? Everything I read in the newspapers is always bad. They say coyotes kill calves, so why don't they ever bother ours?"

It was an interesting point. Why didn't they bother ours? I had a feeling our coyotes liked us. But there were probably good biological reasons for our not being troubled when other ranchers claimed to suffer from depredations. We had to be doing something right on the ranch that other ranchers were not doing. But I was at a loss for ideas. Like Marsha, I'd been fed a ton of negative material on coyotes, but no one seemed to dwell on the obvious good they did.

Nature had given the coyote a serious function, yet I'd never given its place in the scheme of things serious thought or worried how it fit into the grand design. If I planned to defend the animal's presence on my ranch, I owed it to my family to do some objective homework.

Fearing questions I couldn't answer, I left the conversation hanging, abandoned the remains of my breakfast, grabbed a bag lunch from the kitchen counter, and hurried down toward the corrals. As I approached, the old mare nickered to me and came to the fence, hoping for grain. An easy friendship, that one. Undemanding. What did Straight Edge care if I didn't have answers? She had no questions.

AS THE SUMMER PASSED, the coyote seemed to relax his watch over me, though there were few days our paths didn't cross. Sometimes I saw him out on the marshy areas along the Williamson, hunting mice among tussocks covered with tufted hair grass. Always alone, blending with the fading grasses, he would stand statue still, ears focused sharply forward like tiny radar screens, aware of the faintest rustle of grass or squeak from an earthen tunnel. Then, suddenly, the coyote would unwind like a busted watch spring, bound high, and pounce stiff-legged, pinning some hapless rodent to the earth with both forefeet.

If the prey fought back and bit him on the nose, he would give the animal a swift shake and toss, wrinkling his nose in a snarling grimace as he caught it again, this time more carefully. Testing with one paw to determine the rodent's ability to fight back, he might toss it high again. Often the coyote prodded a rodent as though disappointed in its death, as though he wanted a playmate in his lonely, predator's life, rather than a meal. Never did I see him kill with any semblance of anger.

If he was hungry, he ate swiftly, without ceremony. But sometimes after a kill, he would toss the prey aside as though the taste or odor offended him. However, in nature nothing is wasted. Buzzards would descend out of a seemingly empty sky as though they had been spying, and orange and black carrion beetles would appear from nowhere to finish off anything that was missed.

West of the ranch, in the national forest, a series of springs percolating out of volcanic ash blown from Mount Mazama some six thousand years before has, through the centuries, formed a chain of long, wet meadows, or draws, which meander for miles through arid lodgepole and ponderosa pine forests before joining the valley proper.

One day as I was riding the old bay mare across a lonely stringer meadow called Haystack Draw, I saw my friend ahead of me. His head sparkled in sunlight, as though it were a silver object emblazoned with blue. Don Quixote wearing a silver baptismal basin for a helmet. A spectacle right out of Cervantes!

"The Don himself," I mused. "Girded for the fray."

I watched as the coyote spied me and came trotting closer. Something he was carrying in his mouth caught the sun. I was almost prepared to see him packing a lance and broadsword. Grinning at my own bad joke, I named him then and there, "Don Coyote."

The illusion faded as he moved closer and I saw that he was packing a soda pop can in his jaws. Now and again he would pause to toss the can high in the air, catch it, send it tumbling, then pounce on it as though to keep it from escaping to its burrow. Head cocked, he watched it roll down an incline, sprinted after it, captured it, then "killed" it with a shake. Dropping it in the grass, he watched it intently to see if it dared struggle. When it lay silent, he carried it back up the slope, where the whole crazy game began again. Don Quixote tilting against a windmill.

At last, tiring of his game, Don Coyote buried his treasure in the grass and trotted away. But as he was about to leave the meadow, he seemed to remember my presence and scurried back to his digging. Even though the mare must have seemed uncomfortably close, he dug up the can, seized it in his jaws, and scurried off.

Perhaps he intended only to take the can off with him into the forest, but now a new game occurred to him involving me. Once he had moved to a safe distance, he placed the can on a hummock, and, as though to tempt me, he retreated. Pretending no further interest, he appeared to be listening to a conversation between mice. Now and then, however, his quick glance in my direction betrayed him.

Playing his game, I rode toward his treasure, but the Don whirled and raced back, ears close to his head. Scooping up his prize in a wild scramble, he ran off, head held high as though he were delighted with himself. At a safe distance, he skidded to a stop, tossed the can, killed it with a stiff-legged pounce, nudged it toward me with his nose, then crouched behind it, his rear end comically high, tail waving, daring me to try to take it from him. He was toying with me as he had once played at ancient coyote games with his littermates.

Excited to be asked, I dashed forward. My quick rush ahorseback caught him by surprise. He fell over backward in retreat, then fled, looking over his shoulder to see if I rode in pursuit. In triumph, I dismounted, seized the can, and as he stopped at the edge of the timber to look back, shook it at him. Full of wet sand, the can was surprisingly heavy. Teeth had punctured the aluminum and scratched it, as though he had carried it about for some time.

Dropping my reins, I left the old mare to graze and walked toward him, holding the can up for him to see. I tossed it high in the air. The can alit and went spinning, and I went after it, pinning it with my hands in a headlong dive. Don Coyote looked interested and sat on his haunches to watch.

Romping along on hands and knees, I seized the can in my teeth, growled a fierce growl, shook my head savagely, and tossed it high. I lost the can in the sun; then, in falling, it almost took off one of my ears. Another wild pounce, but this time the can spun away, and down I crashed, clutching a twisted wrist.

"You all right, Dad?" Dayton's voice cut though my agony. I sat up, spitting grass seed and pumice, to see my whole brood lined up on horseback watching me.

"Of course I'm all right!" I snapped, clutching one sleeve. "I was just playing capture the can with that coyote over there."

"What coyote is that, Dad?" John asked.

"Right over there," I pointed, looking across the meadow toward the bordering forest. Don Coyote was nowhere to be seen.

I left the pop can sitting on a badger mound in the meadow, and when I rode by the next day it was gone. Either Don Coyote had returned for his treasure, or my children had crept back later and copped the prize as a memento of a good chuckle. Mercifully, the children made no mention of the episode to their mother, and I soon felt at ease again.

For a few days I applied myself to mending ranch fences, but soon curiosity made me venture into the Don's territory to see what new mischief he was about. I found him hunting along the meadows, surrounded by a placid audience of Hereford calves tended by a baby-sitter cow, who treated the coyote as though he were invisible. He was hunting ground squirrels and was not above using the cow as part of his strategy. Ambling haphazardly across the meadow, tongue out as though winded, he waited until the cow lay down and belched up a cud. Using her body for a screen, he rose on his hind legs for a quick look over her back, then, making a swift end run past her tail, caught a ground squirrel out of its burrow. I was so close I could hear crunching sounds as the coyote's strong teeth crushed the squirrel's skull.

For a few moments, as my son Dayton rode up to join me, the Don moved off and sat watching at a distance. I busied myself with a shovel, bolstering up the leaking bank of an irrigation ditch with sod.

"Hey, look!" Dayton said, nudging my elbow. The Don had come back and now came close to the cow, worrying her out of her bed.

He checked the bed for mice that might have taken refuge beneath her. Not finding a meal, he decided to take advantage of the warmth of her bed. It was an act of pure orneriness; he could have picked any spot in the field. As the cow shook her horns at him, he turned around three times, nosed the ground, then plunked himself down on the captured turf, tucked his nose under his tail, and proceeded to sleep. When Dayton and I rode off together half an hour later, he was still dead to the world.

It was often thus. A bit of gentle mischief among the cows perhaps, but he always refrained from killing calves, though I confess my faith in him was not a hundred percent. I was always a little relieved when he trotted off and left them alone.

Often Dayton and I watched together, glad enough to have something like the Don to share. The coyote seemed drawn to cattle, horses, and humans for lack of other playmates, perhaps to satisfy some remnant urges of puppyhood. Often, as the cows lay chewing their cuds, he dashed through them full tilt, using them as racing pylons, delighted when one rose to chase him, but disappointed when they lay complacently and belched up cuds.

Only when baby calves were present did the cows take the bait, one cow rousing the rest of the herd with a dramatic bellow that shouted, "Trouble." Don Coyote loved the grand scene. Crouching just out of reach, facing the action, pink tongue hanging halfway to his knees in a salivating grin, he always managed to dash away just inches from trampling hooves and slashing horns.

"Sonofabitch'll git your calves," a visitor said one day.

"Sonofabitch'd better not," I retorted, angrier at the man for making such a suggestion than I would have been at the Don if I'd caught him in the act. "Hell, in fifty years of raisin' cattle 'round coyotes, we've never had a problem. Rustlers and disease—well that's another story."

"Wal, 'ee'd better stay home," the man threatened ominously. "I ketch'm over your fence 'n' I'll collect his hide fer shore!"

His own ranch was an overgrazed, ground-squirrel-infested mess, and I felt like telling him what he needed around his place was a coyote friend like mine. But I held my tongue. His hatreds ran deep, as had those of his father before him.

My tolerance of other people did not last long, however. One day I was stealing a ride on a snaffle-bit colt, urging it on through the pine forests near the ranch, trying to keep it too busy to think about bucking, and hoping, if it did, I wouldn't fall off right in front of my children. To make sure I didn't, I sent them off to the far end of the field to gather some steers, and I rode off by myself.

As I rode, the colt became increasingly nervous, jumping each little log as though clearing the top of a corral fence and working his ears as though bothered. Suddenly, breaking my concentration, I heard the drone of a small airplane in the distance growing ever louder as it circled the very forest through which I rode.

I had written it off as another Sunday flyer having a
look at my scenery, when, suddenly, it passed directly
overhead. A rain of particles spattered on the ground
around me. Dismounting, I scrutinized the bits of yellow
tallow on the brown, pine-needled floor of the forest.
POISON! The pilot was scattering predator poison on
my private property!

Enraged, I leaped into the saddle and booted the astonished colt into a gallop to intercept the next run of the
plane as it circled and headed back up the valley. Again,
there was a rain of tallow as the plane flew over. Moments later it passed from view, but I had what I wanted,
a clear view of the plane's I.D. numbers on its fuselage.

There was no way I could prevent the senseless slaugh-
ter of animal and bird life, for the lethal doses of the poison, which I assumed was strychnine, were scattered over
hundreds of acres of my forest. Only a few weeks ago I
had aimed a rifle at Don Coyote. Now I was heartsick
that the drop was on the Don's range, and there was little
I could do to save him.

Angry and depressed, I turned my colt's head home-
ward, dreading the fast approach of dusk which would
bring coyotes and other animals from their dens. Soon I
found a Steller's jay lying silent on a bier of pine needles,
looking as though a chunk of summer sky had fallen to
earth.

Not far from the ranch house was a place where my
Lewellyn setter, Spray, was wont to sit and wait for me to
come riding home—patience itself until I appeared—then
dash to me, wriggling and dancing like a communicating
bee as she stole moments alone without competition from
the other dogs.

She had taken up her vigil there for the last time. Now
I found her as she lay writhing in agony in the final, painful moments of what had been a joyous, sheltered life.
Her eyes seemed to plead with me, filled with trust that I
now would help. There was nothing I could do but cra-

dle the poor, confused animal in my arms as she foamed and died and the lights in her eyes blinked out. As I raged helplessly around my land in the next week, I discovered that fourteen coyotes and uncounted numbers of other animals and birds had died from the drop, illegally poisoned in the name of predator control.

Nor did the operation cease with this airborne attack. On a hillside above the ranch, on the eastern shoulder of Taylor Butte, but still on ranch property, I found the hind quarter of a horse, wired to a post. It turned out that the meat was laced with 1080, a savage, nonselective poison for which there is no known antidote, a poison that causes death to the first animal that ingests it and death again on down the chain of scavengers.

Checking with the U.S. Forest Service, I found that the nearest area where the use of poison had been authorized was some twelve miles away to the south, on the Sycan River.

I had no trouble locating the poison plane at the Klamath Falls airport. There it was, sitting all alone on a runway apron as though the other planes had ostracized it. The young pilot was friendly as I took a stool beside him at the airport cafe. He knew me as a cattle rancher, active in local and state livestock affairs, and had no reason to expect that I would be other than friendly.

He disclosed the names of the sheepmen who had hired him. "Just doin' a job that needed doin'," he said. "Those sheepmen been losin' lots of lambs in the area."

I failed miserably at keeping my cool, taking a deep breath to keep from falling apart completely. "There were no sheep permits within twelve miles!" I retorted. "You scattered that damn poison on my private land!"

The young man shrugged and turned back to his coffee. "Got to get those bastards where you find them," he said.

I was ready to start a lawsuit. I tried to get him to testify in court against the persons who had hired him, but he refused.

"Hell," he said. "The hatreds those guys are packin' against coyotes, if I ever opened my mouth in court, they'd fix it I never flew for hire again."

NOT FAR FROM THE tractor, a steer had eaten the root of a wild parsnip plant, and that caustic poison had burned through its innards causing quick death. It may have been that this mountain of ripe good eating so close to his den helped save Don Coyote's life.

Had the coyote lived out on the vast, arid public domain, he might have had to scrounge far harder for a living than on my ranch. Surrounding the knoll where the tractor sat was some of the best soil in the area and an abundance of water. The rich diversity of plant life attracted a rich diversity of animal life, and Don Coyote, tamer than the other coyotes, found himself a niche that had to be Easy Street.

There were plenty of meadow mice and voles tunneling about the hummocks of hair grass, orchard grass, and brome. Deer mice hunted pine seeds close to the rotten logs along the meadow borders. Short-tailed ground squirrels lived in colonies that extended right to the edge of the tractor.

Perhaps, too, my handouts of bologna kept him from wandering out of the area. Don Coyote was the only coyote on my private land to survive the poison drop. Hardly a day went by in which I didn't find bodies of poisoned animals and birds. Except for the Don's spoor near his den, the woods were almost trackless. If we heard coyote music at all, it was distant, like listening to wild geese or sandhill cranes so far away in the heavens as to be invisible.

Being the only coyote about, however, did not make the Don any more dependent upon me. I seemed only a curiosity to keep him amused during moments when he grew bored. If at times he was interested in my antics, curious as to the food potential in what I did, or, as with the pop can on the meadow, playful, there was still a wariness, a caution that dictated that I could come only so close before discretion would move him away. He was like a wave that could come only so far up on the beach before natural forces took it away.

If I was at a safe distance, he would read each corner of the tractor for messages left by visitors, then wash the slate clean with a heavily ammoniated message for any ranch dog fool enough to trespass. Each corner required equal drenching—north, east, south, west. Drifting down the hillside, the fumes made me cough a hundred feet away. Science has yet to explain why, on similar diets, wolf urine is only mildly fragrant compared to its smaller cousin's, the coyote's.

As a final part of the ritual, the Don would raise his hackles, glance about as though some challenger to his territory were drawing nigh, then tear up yards of turf with powerful strokes of his hind feet.

The Don never hurried. Once he had restaked his territory, he scratched, plunking down on his tail to do the side of his neck, then balancing awkwardly on three legs to do his rib cage. Often he fell over in the process, and once down, chewed his flank or the root of his tail into a white froth of saliva as though to drown his fleas.

It may be that before a coyote dines, he prefers not only to look right but to smell right. Instinctively, the cattle avoided the deteriorating remains of the steer, sometimes bellowing their fear and charging up to sniff at the lost brother, but avoiding grazing near him. Nurtured by such a concentrate of natural fertilizers, timothy and bluegrass grew tall from a supportive base of bones, effluvia, and discarded maggot overcoats. The odor would have sent a bronze Remington cowboy clutching for his bandana, but to the Don it was pure heaven.

Here he would pause to roll, placing one pale cheek on the pile while he wriggled along on his ribs. That area completed, he did the other side, then, in turn, his back and stomach, as though the stench were a powerful insecticide and he were determined to leave no flea ungassed. Only when this elaborate toilet was completed did he saunter over to the rock to see what the ranch kitchen had provided.

For an animal living off the land, adaptable to varied diets, and a great opportunist who in lean times no doubt accepted a good many lesser foods just because they were available, when it came to my wife's preparations he could be mighty finicky. Bologna sandwiches he gulped down as though he were the runt of a big litter. But offer him peanut butter and jelly, and the ruff of fur along his back rose, his nose wrinkled showing fangs, and his yellow eyes crackled with anger. Glaring at me with disgust, up would come his hind leg until the foot pointed almost skyward, drowning the sandwich in a yellow deluge. Kicking dirt over the rock with powerful, staccato rips of his hind feet, off he would trot without a backward glance at me.

His resentment might last a week before he was tempted to notice me again. Day by day my offerings piled up on the rock. When he was ready to forgive and eat, I knew immediately, for he would deign to notice me again.

Even though I was his vassal and he used my tractor as a hotel, message center, and urinal, he did nothing that even remotely showed one shard of affection for me, offering no sign that he valued my company above that of the tractor, the old mare, or, at times, the cows grazing in the pasture. And I guess I needed to be liked. Once, just once, I thought I detected a faint wag of that ever-so-straight tail, a faint grin on that ever-so-impassive face, but both swiftly disappeared, as though he had caught the impulse just in time.

Only by accident did I break the barrier between us. It was September, stretching toward round-up time, when my dereliction of ranch duties began to catch up with me, and all those uninteresting tasks I'd neglected during the summer became emergencies.

Yearling cattle had scratched and rubbed themselves through sagging, untended fences until there were as many heifers among the steers as steers among the heifers. It took long hard days on the part of Gerdi, myself, and the children to separate them in time for the fall roundup, when we customarily gather the grass-fat cattle and ship them off to feedlots in large aluminum cattle trucks.

Having so much to do, I could ill afford spending hours in the field observing a coyote, especially one that, after three months of devoted catering on my part, could not reward me with so much as a wag of his tail. What did I want with this arrogant brush wolf that haunted my land? Not to dominate him. All I wanted was to be liked, to be accepted as a friend so that I could observe him better and learn to understand wild coyotes and their ways.

When I finally managed to visit the coyote den once more, the tractor seemed lonely and deserted. I stood on the rusting tracks and peered below. Beneath Big Alice I spied a labyrinth of tunnels, but spiders had built webs across most of them. In the one likely tunnel, a huge toad with ruby eyes squatted like some Japanese sumo wrestler. What coyote tracks I could see in the gloom looked anything but fresh.

Despairing of ever seeing Don Coyote again, I sat down on the sandwich rock to think. Straight Edge had just lowered her head to graze when she raised it again to look past me. I heard a whine and turned to see the Don sitting not twenty feet behind me, head cocked as though wondering what I was about.

Carried away with excitement, I tipped back my head and uttered my very best howl.

For a moment the coyote stared at me, uncertain whether to stay or flee. Then, suddenly, his tail began to wag, he pawed the air with one front foot as I had seen him do to others of his kind, wriggled in excitement, laid back his ears and grinned, then tipped back his head until it almost touched his back, and erupted in pure joy.

It didn't seem to matter a damn to him that my howl was wavering and awkward, an act of puberty, while his was a marvelous, hysterical, bursting anthem to freedom, the far off mating season, wilderness, and the chase. Perhaps to him I was a pup, voicing my first tentative, wavering, and uncertain song to my littermates.

But at that moment an amazing transformation took place in our relationship. Because I howled with him, Don Coyote took me for a friend. For the first time ever, I got the tail wag my inherent loneliness had been demanding.

5

AUTUMN CAME AND WITH it that frenzied and—to the landowner—often frightening event known as deer season. Each year there are a lot of good people out there in the woods who all year long have worked hard at their professions and this is their respite. They seek to get away from it—the jobs, the kids, the camper payments—and go off for a few days into the lovely lonely forest, breathe fresh mountain air, shed a few pounds of fat, relax with good company before a campfire, watch nature, and remember how much better things were when they were kids. They take their good manners with them, treat ranchers and farmers as human beings a good deal brighter than those portrayed on television, are careful with fires, leave gates the way they found them, look before they shoot, leave a clean camp, and have a great, enthusiastic time whether or not they see game.

But there are also some real clods out there, a tiny mi-

nority who make it a nightmare for everyone, especially
that farmer or rancher who must go about his daily business as though there were no war in the woods.

Although our ranch is well posted, there is always a
danger that the wrong hunter will show up on it and ignore the signs. The national forest lands that surround us
can be classified amongst the most lawless in the state.
Although the Forest Service works at attracting the public, it does little or nothing to see that they behave once
they are there. And the state police have enough to do
patrolling the settled areas and highways, leaving the rest
of that vast area to chance.

Trouble started two days before the season when Gerdi
caught a camera buff who had parked his car off the road
to town, shot our old black Percheron draft mare, Bess,
and set up a camera blind beside her to photograph tur-
key vultures.

I knew that Bess's spirit must be off somewhere up to
its knees in grass, in whatever heaven there is for a horse,
who had toiled faithfully many a year pulling our wagon-
loads of hay through knee-deep snow to hungry cattle.
She would not mind if I made use of her earthly remains.
As her mate, Babe, knickered for her through the woods,
I drove the tractor out and dragged Bess's remains away,
wishing as I did so that I had been the one who caught
the trespasser and not my gentle wife.

As I left the carcass in a thicket near the tractor, I saw
the Don, nose to the wind, working his way toward us. I
gave Bess one last pat on the neck and left, glad that her
death was not a complete waste and that she might do me
one last favor in keeping the Don well fed and content,
away from the borders of the ranch where he might get
shot by hunters.

I was tacking up more No Trespassing signs along one
of my fences that was skirted by a public road, when the
sheepman I suspected of putting out the 1080 poison station on our hillside drove by on his way out to his bands

and paused to give me some advice. On the seat beside
him was a large kettle of chunked meat and tallow, and it
did not take much imagination to guess what he'd been
about.

"Still got thet goddam coyote aroun'?" he asked,
shooting a lip load of snoose just over the edge of the
rolled down window. He didn't wait for an answer.
"Wait till calvin' season. Kill your calves, thet's whut. 'N'
deer. Coyotes git thick roun' here, kill ev'ry goddam
deer in th' country. Shoot thet sonofabitch, thet's whut
I'd do. 'Ere, use summa this 'ere poison. I gotta kid's
pretty sharp wi' a rahful; let 'im on your place 'n' he'll
mek short work o' thet wild dawg, thet's fer shore."

I let his comments carom off my silence and wished I
knew more positive things I could say in Don Coyote's
defense. Such facts would not ease his hatreds, but ex-
pounding them might make me feel less helpless. Filled
with a hopeless sort of anger against such men, I went
back to nailing up signs, noticing with surprise some
minutes later that I had smashed my finger, and blood
was soaking through my glove.

That afternoon Don Coyote only sniffed at his bologna
sandwich on the rock, indicating that he was well fed. We
had a good howl together before we shoved off down the
ranch. Fall was in the air, the aspen leaves trembling
golden on their petioles, and the sun brassy hot. The old
mare pranced like a colt, seeing ghosts behind every log,
giving me a gut ache as though I'd just eaten a peck of
green apples.

"Slow down, you old rip," I grumbled, "or I'll hitch
you up to a farm wagon."

The ground squirrels had long since taken to their burrows for the winter, and now a coyote's take was mostly
meadow mice, pocket gophers, and an occasional frost-
stupid grasshopper.

Don Coyote trotted along behind my mare, but there
was a nervousness about the animal which may have been

only a reflection of my own. But I wondered if he could sense that tomorrow the lead would be flying thick and fast, and that one played roulette with death just by leaving the stone house. Come to think of it, even the stone house wasn't safe, for in my desk drawer I keep a bullet that had ricocheted off the roof of my station wagon and on into my daughter Marsha's bedroom.

When I left the coyote that night at dusk, I was relieved that he went trotting off toward Bess's remains.

"Sure hope your coyote friend doesn't get shot," John said that night at supper. "Tomorrow morning I'll ride with you ahorseback to patrol the ranch."

"Better take the pick-up," I suggested. "It's safer." I patted him on the back, glad that he was man enough to care.

That night I tossed and turned in my bed. I kept hearing shots, pictured my friend dead and hanging by his hind legs from a barbed wire fence, a classic barbarism of coyote haters. At dawn I pushed myself up on the edge of the bed, exhausted and ready to go back to sleep. It seemed so unfair that I had to leave that warm bed to protect my land when there were plenty of No Trespassing signs. Daylight found me saddled up and riding, heading for the tractor.

To my relief, Don Coyote was at home. He came out of his lair sleepily, stretched, and regarded me curiously, as though his inner clock told him I was early. Pulling out a whole ring of bologna I'd stolen from the pantry, I offered it to him. He leaped high for it and trotted off, but evidently he had eaten to satiation, for he dug a hole in the sand, dropped in the meat, then pushed dirt around it with his nose. The wet sand clung to his face almost to his eyes.

For a moment he stood beside the tractor, looking at me over his shoulder. The early morning sun accented the reddish orange fur on his ruff; his thick pelage, rippling in the breeze, gave indication that an early winter was in the offing. For a moment he sat and scratched, eyes half closed in ecstasy, then, yawning, he rose to his feet and disappeared into his den.

I slipped away, hoping he would stay out of harm's way. Already I could hear the first shots of the deer season echoing from Taylor Butte and the long ridge that slopes south from Yamsay Mountain. The shots were a long way from my property, but my jaws clamped together in an unreasonable anger I'd never felt before.

Straight Edge tiptoed as though death lurked ahead, making bears out of stumps and rocks. No sooner had she laid one lurking monster to rest, than she picked out another half a mile ahead, stared, snorted, pranced sideways like a crab, and sweated buckets until she got by that one without being eaten.

Perhaps the old mare had been with me so long she sensed my own tensions. I hated the woods today. I kept waiting for the crash of bullets to spin me about. A German sniper in the war had taught me that the sound comes later, after you are hit. To let hunters know my whereabouts, I whistled old familiar tunes until I sickened of them. My lips and mouth seemed a dried up water hole, trampled by a thousand buffalo.

I saw my first hunter poised outside my fence. He watched me through his scope, either trying to put antlers on my head or see what manner of man lived way out here.

Unused to being aimed at for any purpose whatsoever, I cursed aloud, a deep, justifiable anger burning within. How did I know he wouldn't bump the trigger by mistake, that he wasn't mentally unstable and the temptation to shoot another human too much to resist? I put a grove of pines between us and trotted on my way.

It struck me that the situation into which Don Coyote had fallen was of my making. I had courted his trust and done my best to tame him. Had I really valued him as an animal, the best thing I could have done for him was to shoot at him, throwing dirt up behind him every time I saw him and thus condition him to fear Man. But the first shot would have driven the animal off my property, where he would have had even less chance of survival. Here, at least, he was cherished and protected. His den was in the middle of our six thousand acres. If only he would stay close to Big Alice, he would be safe.

The hills popped with gunfire. The deer had been roused from the thickets and were fleeing toward the sanctuary of the ranch in blind panic. I hadn't heard that much shooting since the Battle of the Bulge.

A pick-up truck was coming, moving too fast down the road. It looked as though it had no driver, then I saw the top of John's head just peering over the dash.

"Dad!" he blurted, braking to a stop. "Somebody left a gate open and about twenty of our heifers have gone exploring. They're heading up a draw right toward the firing line!"

There was no time to visit. Straight Edge forgot her years and lumbered into a gallop. Charging out the open gate, I picked up the trail in the pumice dust and pushed on. Dead branches broke as I passed, and brush scraped my chaps. I located the sightseeing yearlings kegged up in a lodgepole pine thicket, circled them, shouted them into motion, and headed them back down toward the ranch.

They were streaming through the gate ahead of me when I heard a coyote bark, followed by a long, joyous howl. There in the middle of the open meadow sat Don Coyote, watching my approach.

Quickly I glanced about. No hunters in sight! If I hurried I could lead him back toward the heart of the ranch before hunters outside my fence spotted him and fired.

Kicking my horse forward, I drove the last animal through and leaned from the saddle to close the gate. Though winded, the old mare sensed the hurry I was in and shot forward at the click of the gate latch. As I charged ahead, I could still see Don Coyote, sitting exposed to the hostile world across my fence, waiting patiently for me to join him. As I headed toward him he stood up, pawed the air at me, wagged his tail, grinned showing a white-toothed smile, and tipped back his head to howl.

Suddenly Don Coyote staggered backward as though stung by a hornet, leaped high in the air, and fell, snarling and biting himself as he rolled on my meadow. A shot roared in my ears, and the report caromed off the wall of the pine trees. From the timber behind the Don, a trespasser rushed toward the fallen animal, so excited over his kill that he failed to hear the beat of hooves, or my cries of rage and anguish. Jerking a knife from a sheath on his belt, he knelt, and as Don Coyote bit blindly at a clump of grass and was still, the man slashed off his tail.

Speechless with rage, I skidded my horse to a stop.

"Lookee thar!" he laughed, holding up the plume for me to see. "Sure drilled that bastard! Sure fixed his clock!"

I tried to keep my voice from breaking, to keep from stepping off my mare into midair to do violence. "That makes you a goddam big hero, doesn't it! You just shot my tame coyote! Now take your fancy gun and get the hell off my land!" My voice went through all the ranges of puberty as I tried to keep it under control. How very close I was to killing him!

For a moment the hunter bristled, glaring back at me, fingering his weapon as though considering its use. "You damn rich ranchers are all the same," he snapped. "Want all the game for yourself! They ought to fix it so's you can't own land and keep folks off!"

I tried to make my voice come, but all I got was a squeak. I wanted to ask him where he knew of any public land that was better managed for wildlife than mine, or where wildlife would go for privacy if private lands were made public. I wanted to tell him just how handy coyotes were to have around to clean up carrion, to eat the rodents that devoured my grass. I wanted to tell him just how important predators are to the hunter in cleaning out the sick, the lame, and the lazy from the genetic pool. Instead, all I could do was turn, sick at heart and stomach, and get out of there.

I could bear only one last glance at the animal that had come to mean so much to me. The bullet had amputated a hind foot, and the blood made a dark medallion on the light tan of the autumn grass. His teeth showed white in one last distasteful grimace, and a light breeze stroked his lush fall coat as though the hands of spirits were already there to stroke and comfort him.

I rode off without looking back, following along behind the hunter to make sure he left the property. I could not bring myself to go back and bury the animal. Let him return to dust the way Nature returns all her children except Man to their borrowed elements. I had already subjected Don Coyote to enough.

At supper I sat silent as Grant's tomb, and my family waited patiently for my anger to head. "A hunter shot my coyote friend," I blurted out suddenly, fighting for control. "The guy was right on our land. A few seconds earlier and I could have saved him. The Don was sitting in the meadow waiting for me when that trespasser cut him down."

My family stared in shocked silence into their soup plates, feeling perhaps that a bit of Yamsi history had ended, each, I'm sure, treasuring his own favorite memory of the Don.

ON A HILLSIDE AT the south end of the ranch lies the huge, partially dismembered carcass of a pine tree surrounded, as in memoriam, by a cathedral grove of healthy children. Like Don Coyote, it too had once been territorial. Its great roots had thrust out through pumice sands, brooking no rivals to its small domain, soaking up nutrients and moisture, scattering seed over the land but crowding out its progeny when they upstarted too close to its system.

Then one spring, sensing its own death, it had expended its last reserve of strength to produce a mast year, showering the ground beneath with seed which, on germinating, grew thick as coyote hair and took over the area that the great roots had once dominated.

The great old monarch pine had fallen one year in a winter storm and soon, softening with age, had become my worry log, a place where since boyhood I have sought refuge to think out problems. With Don Coyote's death I found new need for solace. Riding past the grove the next morning, I left Straight Edge to graze on the meadow while I walked up through the pines and sat myself down on a spot already worn by years of my meditation.

I suppose a wiser man would have put coyotes out of his mind forever and acquired new interests, but that wouldn't have worked for me. With Don Coyote gone I was still troubled by a sense of predestination—that perhaps I had been programmed even as a child to become involved with coyotes and that, far from being through with them now, our relationship was just beginning. Like many people I had run slipshod through life without really taking time to understand who I was, why I did things the way I did them, or what road I was destined to travel. The Don's death had kicked the joy out of me, and now I sat comfortably embraced by the roots and thought back on the past.

God knows I never even intended to be a rancher. It just happened. There was a time in my life when the wheel was still spinning, a time when my possessions could have fit into a shoebox and my world was a small northern Michigan town on Lake Superior and a lonely summer cabin on a nearby wilderness lake. Had it not been for the arrival in town of my cattle rancher uncle from Oregon after an absence of twenty-five years, I would be there still, gray-haired now, sitting on the back bench of Richard's Sport Shop, talking about the taking of partridge and woodcock with the other old-timers.

My uncle stood tall, red-haired, and slim before the hearth at my grandmother's house, part of his enormous presence hanging with his big Stetson hat on the coatrack in the vestibule, but retaining enough and more to tongue-tie the Hyde children peering at him from various corners of the parlor. He and my mother had always been close, and now it was as though he had come back to pick one of her litter of children to ease her burden.

That we children were a drain on her was not hard to guess. The day the stock market crashed in 1929, my father suffered a reversal in his fight against what came to be known as multiple sclerosis and had to give up his job as Superintendant of Lumbering for Cleveland Cliffs Iron Company in northern Michigan.

We kids came home from school one day to find him wedged helplessly where he had fallen between the back porch banister and the house. It took the combined effort of neighboring cooks and handymen to extract that two-hundred-pound man and put him to bed.

He was up the next day still trying. He cut himself a cane of moosewood and continued to walk twenty miles a day through the forests that he loved. Whatever curse afflicted him, he felt that good food, fresh air, and lots of exercise could bring about a cure.

Nothing helped. When, finally, he was downed for good, his loggers, though hard hit by the Depression, walked the sixty-odd miles to town packing him gifts—a wild berry pie, a bracket fungus inscribed with a poem or prayer—and love.

In those days I dreamed of becoming a naturalist. Many a summer morning at our camp on the lake, I'd creep to my father's bed to wake him. We'd lie listening to the sounds of the forest, and he taught me to identify every instrument in that symphony.

"What's that?" he would ask, testing me, and I would proudly identify an ovenbird, cedar waxwing, or kingfisher, until there wasn't a creature in the forest I didn't know as a friend.

There is a built-in solitariness that accompanies that sort of interest. Boys my age were into sports, Model A Fords, hunting animals instead of making friends with them. More and more, I felt cut off from the gang, choosing to wander the woods alone.

One summer day when we were in town for groceries, I saw a group of my schoolmates on a street corner, clustered about a couple of the current athletic heroes. A Model A roadster went chuckling past, top down, a pretty blond girl snuggling close to the driver. The boys on the corner stared enthralled, each lost in his own particular fantasy. I hadn't seen a boy my own age in weeks; how desperately I wanted to share my experiences.

"Hey, you guys," I said bounding up to the group like a setter pup. "Guess what I found this summer! A Blackburnian warbler's nest in a hemlock tree!"

I can still feel their stares as each gave me the silence of his contempt and turned back to watch the retreating Ford.

"Five eggs," I said weakly, still trying, but looking about for a tree behind which I could hide.

That evening, back at my favorite glade at the lake, I tried to make everything fit. A ruffed grouse drummed a muffled heartbeat in the distance, and a wood mouse came dancing from a rotten log to take a pine seed from my fingers. "I don't care," I said to myself. "I just don't care."

In front of the screened porch where my father sat out his long summer days in his wheelchair, a small cedar had been spared the axe and grown up straight and lush. One year it hardly reached my ears, by the next it had cleared the eaves. During hot afternoons my dad jockeyed his chair along the porch to follow its friendly shade.

How great was his delight when he discovered a pair of cedar waxwings building a nest in the cedar only a few feet from where he sat.

"I know! I've seen it!" I replied sulkily, as my father pointed it out. I was ashamed of my interest in nature. I stalked off into the woods alone and did not return until pines, birches, maples, and cedars were all melded together by darkness and tinder wood gleamed in the damp bottoms.

Two days later the waxwings finished the nest; five days later the female began to incubate in earnest, while the male brought her choice morsels or sighed his soft music from the pines along the lake.

I pretended disinterest; if I came on the porch at all it was because there was something my father needed and my mother was too busy to take care of.

"They've hatched!" my father exclaimed one day, two weeks later.

Indeed, there was great activity about the little cedar. Both parents flitted about, beaks filled with damsel flies and other summer insects. Just above the lip of the nest, I could see tiny yellow beaks, scarlet mouths, and fuzzy heads on thin, uncertain necks, straining each for a share.

"What's hatched? Oh, yeah. The birds. Well, I gotta go. See ya."

I tramped through a cedar swamp and sat down in a little sphagnum glade I loved. Picking a yellow violet, I spun it between my fingers like a propeller, watching as it became a blur of gold. I had no idea of the drama that was unfolding beside my father's porch.

His first inkling of danger came when the cedar waxwings began to scold. In a moment they were joined by a pair of chipping sparrows, then a shrill-voiced robin. The surrounding forest echoed their distress. My father peered up at the nest, but all seemed calm; the young had just been fed, the mutes carried away by the parents, and no errant heads showed above the lip of the nest.

When the scolding grew more frantic, my father wheeled to the edge of the porch for a closer look. At the base of the tree, motion caught his glance. A huge black and yellow garter snake was climbing the tree, intent upon robbing the nest.

For a moment the snake paused, round, black eyes glittering, red tongue darting to read its prey. My father shouted to my mother for help, but she had gone to gather berries for a pie. Again the snake began to move, draping over one limb after another, sliding round and round up the tree. The parent birds were in a frenzy now, flying as close as they dared. When my dad yelled at the snake, it only seemed to hurry.

Wheeling his chair to the screen door, Dad lurched forward, tearing the door off its hinges as he fell.

When I returned to camp, the surrounding woods were quiet. I found my father lying on his back, looking up into the tree. On one cheek a scratch still trickled blood; his torn clothes were a sight of pine needles, dirt, and last year's blueberry leaves. Over the lip of the nest, I could see the mother waxwing, crest swept back in peace as she warmed her brood.

"A snake!" the old man grinned in triumph. "Biggest darned garter snake you ever saw. He was after the babies in the nest; had to climb halfway up that tree to get him down. Grabbed him by the tail just as the branch broke. But I saved the nest." He chuckled shyly. "Can't say as much for that harmless old snake. I guess I fell on him."

"No foolin'!" The boys on the street corner faded from my mind. Dad weighed nearly twice what I did, but I managed to roll him off that flat snake, brush him off, and get him back in his chair before my mother returned.

Watching that brood together, we learned a lot about cedar waxwings, my dad and I. One afternoon we were playing a game of cribbage on the porch when the young made their first flight, deserting the little cedar tree forever as they beat heavily into the pines. The nest looked suddenly empty and forlorn. When he added his points, my dad took advantage of my lack of concentration and moved his peg backward instead of forward.

"You won," he said.

"No," I argued. "You won!" He'd never thrown me a game before. Suddenly I knew that he was thinking that soon I too would leave the nest, and when that time came I would have his blessing.

A few days later a letter arrived from my uncle brag-

ging that he could step out the front door of his ranch
house with a dish pan and scoop up enough trout for
breakfast from his house spring, and that at night the
coyotes made the forest ring with endless song! Now
what kind of a letter was that to write to a boy if you
didn't expect to see him at your door?

7

THE WORRY LOG SEEMED to sag a little under my weight and I was reminded just how many years since my youth that old ruin had lain there. I let my mind drift back to the time I had first come to Oregon and seen my first coyote.

I lay on musty hay in the pitch black of the great A-frame barn and listened to brand new sounds. I was so fresh to that world it was like being born again. Fourteen years old, a fresh, bare-root transplant from the wilds of Michigan to the wilds of Oregon, the youngest by ten years on my uncle's rough, tough cattle operation. In the lingo of the Oregon ranch country, I was a "button."

Wide awake in my excitement, I heard the mountain winds blowing in off the vast, prairie grasslands of Klamath Marsh, the creaking of hand-hewn lodgepole pine rafters, the contented snuffling and gristmill tooth-

grinding of horses as they munched a fragrance of timothy, bluegrass and clover hay from their pole mangers,
and the rasping, buzzing snoring of the crew, eighteen
cowboys and hangers-on, all strangers to me.

It was the twenty-fifth of June, 1939, yet outside the
barn a foot of fresh snow had come down off the Cascade
Mountains in a freak storm. Under the thin blankets of
my summer bedroll, I shivered, my body aching for
morning. Now and then a bullet of sleet made a direct hit
on a crack in the shake roof and caromed off my cheek; I
drew my knobby knees tight against my chest for
warmth and wondered how I had ever traded home and
family for this.

I had just begun to doze when, suddenly, from just
outside the barn came weird, ghostly, insane screams. I
sat up straight. "Help!" I shouted. "Wake up, everybody!
Somebody's hurt out there!" Leaping from my bedroll, I
trampled across a dozen bodies in their beds as I rushed
across the barn.

A sudden orange glow of a match stabbed the darkness
as the old foreman, Homer Smith, lit a kerosene lantern
and held it high. Short and stocky in his nightshirt, he
looked like a bull in a pup tent. "What the hell's goin' on,
kid?" he bellowed.

Like pale ghosts in their long-johns, cowboys charged
from their beds as though the barn were burning.

"Listen! Outside!" I gasped. "Hear that?"

Inside the barn there was silence as everyone hushed.
There it was again! More screams, then the laughter of
maniacs.

"There! You hear?" I hissed, proud to be of service to
my elders, and for the first time in my life finding myself
in the limelight.

The gnarled old cow boss rubbed one hand over his
grizzled jaw in disbelief.

"Hell, that's nothin' but a pack of coyotes; reckon
yuh'd holler too ef yuh had ta set with your balls in the

snow. Where'd yuh hail from, yuh never heard coyotes howl before?"

My face grew hot. "Coyotes?"

"Damn a kid around a ranch," a cowboy muttered, trying to straighten out the ruin of his bedroll.

The light from the lantern went out abruptly, and the night hid my shame.

At dawn I arose, trying to be quiet. Here and there a horse looked up from his manger and tried his rope, knowing me for a stranger. I stepped over sleeping bodies and moved out the barn door into a world of snow. At 4,600 feet, the thin, damp mountain air chilled me through my stiff, new denim jacket. For a moment I leaned against the weathered wall where a spot of early morning sunshine bathed the unpainted barn boards with a yellow glow.

The wind loosed a slop of wet snow from a pine branch high overhead. Already the unseasonable snow was melting fast. On the east side of the barn, the eaves loosed a larger and larger torrent which bored holes in the snow and turned the pristine white into a frothy, brown slush.

Not a hundred feet from the barn I found tracks, spaniel-sized, where two coyotes had sat on their haunches to howl, then gone off more or less together, tracks crossing here and there like climbing grapevines, wandering from one emergent grass clump to another as though sniffing out mice.

Despairing of seeing them, I was about to turn back when I caught movement in the snow. They had been lying half immersed in the fluff, watching my approach; now they stood up, moved part way toward the safety of the timber as though improving the odds for escape, then stopped to observe me further.

What a disappointment! As the famed desert philosopher Reub Long said when as a boy he first saw the Pacific Ocean: "I thought it would be bigger!"

I had expected wolves at least. Yet here they were, small, tan, dog-like creatures with erect pointed ears, sharp fox-like muzzles, spindly legs, and long, bushy tails. Still shedding their winter coats, they looked as tattered and moth-eaten as an unsold fur coat after a Salvation Army thrift sale.

I eased closer for a better look, but they became nervous and loped for the timber, tails tucked between their hind legs as though guilty of mischief. Safe amidst a welter of fallen lodgepole pines at timber's edge, they stopped to lift their noses and search for my scent. A bell clanged loud and clear from the cook-house porch, and they fled as though dodging bullets.

As I turned to head back to the ranch buildings, I saw a line of cowboys drifting out of the barn, heading for chow. I slowed, not quite sure whether the grub would be worth the razzing I might take, but the men ignored my presence. Still lost in dreams, they stood hunched against the cold. Half a dozen at a time, they washed their faces in tin basins along the cook-house wall, throwing the milky suds steaming into the snow.

Standing last in line as befit my station, I saw a flash of black and orange as a western tanager flew into a lodgepole pine right over our heads. It was the first I'd ever seen, and I longed to point it out to someone, but I held my tongue. I was learning fast.

For three weeks we gathered cattle from various sections of the Marsh and branded four to five hundred calves a day. It was dawn to dark work. As the lone button on the outfit, my job was to push the kicking, struggling calves up a plank chute onto a branding table—an iron maiden that grasped the calves securely, sandwiching them in steel, and turned them on their sides for branding, vaccinating, dehorning, earmarking, and castrating. To atone for my early disgrace, I worked like a demon.

At night I crawled into my bedroll exhausted, shins a welter of bruises, for none of the cowboys had shown me

how to crowd close to the calves and let them kick past
me. Had a pack of coyotes howled in my ear, I would not
have wakened.

Being the only kid in the outfit, it was natural that I
take a certain amount of abuse from the crew. The men
did show some respect for how hard I drove myself, but
seemed determined to take revenge for my waking them
over a couple of coyotes. Homer Smith issued me a tall,
active dapple gray with a snort like a carbide cannon, and
when I objected that the animal might be a little too
much horse for a guy who three weeks before had never
been within touching distance of one, the old foreman
shrugged with a merry wink at the rest.

"Hell, kid. Thet hoss is plumb gentle. Why, he nursed
his mother when she was sick."

The crew livened up their dull moments by riding behind my green horse, trailing reatas or jackets beneath his
tail just to see me forked end up.

I spent a lot of time on the ground, but as I didn't have
much choice, I kept crawling back on. At length, an old
cowboy named Ash Morrow found pity for me and took
charge.

"You got a definite problem, boy. Thet hoss is a-
learnin' to buck faster than you're learnin' to ride. Either
you got to git better at ridin' or get plumb good at fallin',
one of the two."

Ash seemed to sense my gratitude and leaned back in
his saddle. "Here," he said. "Git your feet forward and
lean back like this. And think positive for once. Tell
yourself next time you're goin' to give that fool hoss a
ride!"

That night I did a lot of positive thinking, but it didn't
help. The very next morning the horse tested my resolve
by bucking me off against an iron water trough, badly
bruising my ribs.

"Hell, thet's much better," Ash said as I lay there hoping to die. "Thet's three jumps longer than Ah ever did see yuh stick!"

Preoccupied with watching my horse's ears to predict his evil intentions, I got only an occasional look at coyotes on the Marsh, and those generally as I was afoot, chasing my horse back to the corral. The animals stayed in the distance, keeping alert for cowboys, who often roped at them for sport.

One afternoon Ash Morrow and I ate our sack lunches together in an aspen grove beside the Great Northern Railroad track, letting our resting horses graze as we munched on chaws of salty venison jerky.

"Yuh want to see a coyote up close, kid? Jes watch along the track as this next train goes by."

The freight train was almost upon us when I saw a coyote dash from the thickets and crouch beside the tracks like a dog getting ready to chase a car. In full molt, he was no beauty, but I stared at him in fascination. As the train thundered close beside him, wind fanned a ruff of fur on his neck. Ahead of the train short-tailed ground squirrels played their games of chicken, standing tall, then at the last moment dashing for their holes.

The coyote stayed low until the train roared by, then ran after it in hot pursuit.

"What would he do with the train if he caught it?" I giggled, amused by the whole antic.

"Hell!" Ash snorted. "He ain't after the train! It's the squirrels. When a train comes by, sometimes those fool squirrels charge down the wrong hole. Once the ground stops shakin', the mister of that hole runs the stranger back out, and the coyote's right there to catch himself a meal."

Sure enough, just as the coyote came to a hole, a fat squirrel came running out, pursued by the owner, and the wild dog made short work of them both. Within a minute, before the train had rounded the next bend, the coyote had killed two more.

"Ah call him 'The Railroader,'" Ash said. "He's been workin' thet game long ez Ah c'n remember."

The coyote devoured two of the squirrels on the spot and had just carried off the third to consume it at his leisure, when suddenly he sensed our presence. Instantly, where there had been a coyote there were only shadows. He vanished into the lodgepole and aspen thickets, leaving only a couple of dead squirrels to mark his passing.

Often during those weeks on the Marsh, I sat in the thickets and watched The Railroader ply his trade. Sometimes he caught one squirrel, sometimes three, but his ploy always worked.

I never saw him with another coyote. When I mentioned that to Ash, he looked a bit wistful.

"Hell," he said. "Ah reckon he's jes like me. When it come to thinkin' 'bout family duties, Ah jes drifted on."

Sharing an interest such as The Railroader, Ash and I became friends. He was along when my horse spooked at a badger, bucked off through a mess of windfalls with me, and I managed to ride.

"Son," he said, grinning as though he'd made the ride himself. "Ah shore don't want tuh spoil yuh none, but there's hope. Yessir, one of those days yuh jes might make a hand."

FROM MY SEAT ON the old worry log, I saw motion through the trees and my heart leaped. For the moment I forgot that Don Coyote was dead, and unless I was lucky enough to go to coyote heaven I'd never see my friend again. The motion turned out to be an old doe heading down to water with her tongue hanging out. Now and again she would stop and look back over her shoulder to make sure nothing pursued her. She was bone thin, and a scar across her hip showed where a bullet had almost missed. The wound showed signs of healing, but she was in poor shape to survive the winter.

She soon lost herself in the tall grasses along an irrigation bank, and I began to reminisce again, this time about the day I left the Marsh and saw, for the first time, the ranch that was to be my lifetime home.

★   ★   ★

Coming west, I had looked forward to seeing my uncle again, but it was some weeks before I ran into him. By then my boyhood fat had melted off, never to return. The word of my coming must have upset my uncle's bachelor existence as much as the arrival in town of an illegitimate child, for he left orders for someone else to buy me boots, Levis, gloves, denim shirts, jacket, and Stetson hat. And it was one of the hands who packed me off to the buckaroo camp on Klamath Marsh, challenging Homer Smith to "make a man of him if you can."

At the time, my uncle owned over twenty thousand acres of land, and ran six thousand head of fine Hereford cows on four ranches and a hundred fifty thousand acres of range. His headquarters was the ranch Yamsi; he also maintained two summer ranches—the Klamath Marsh and Calimus—and a winter ranch—the BK, in the Bly Valley.

Working for such an outfit, cowboys went where the work developed, and we could never be quite sure in the morning which floor our bedrolls would hit at nightfall.

I had just become accustomed to life on the Marsh and was enjoying my visits with The Railroader, when Homer issued me an even tougher bronc than before and ordered me off cross-country to the headquarters ranch, Yamsi.

"You, kid!" he said, packing a fingerful of Copenhagen snuff under his lip. "Take this old hoss and head east 'bout sixty miles toward that mountain over there. You're in luck. The old man wants yuh ta learn hayin'."

I eyed the horse dubiously. "What kind of a killer did you come up with this time?"

Homer grinned. "His name's Sleepy. Hell, yuh c'n handle him good. Ah'll ear 'im down in the corral, then open the gate fer yuh. Better pee afore yuh go, 'cause he won't let yuh back on. By the time he's made thet sixty-mile ride to headquarters he'll be plumb gentle."

"How will I know the place?"

"Cain't miss it. Other side of thet long ridge, you'll come to a river. Jes follow it on to the head."

I tied my bedroll on behind my saddle, but the big bay blew his nose and busted his tie rope every time I laid a hand on him, trying to kick my ear off.

"Maybe he wants me to walk with him instead of riding," I said hopefully.

Homer blindfolded the horse to get me mounted and turned him out of the corral. "Keep him busy travelin', kid," he shouted. "That way he won't git no time ta think!"

Away I went riding a horse I couldn't manage, to a ranch I'd never seen, owned by an uncle I scarcely knew, to a destiny I could not have dreamed of.

I got along fairly well until about noon. I was getting pretty desperate to get off when the horse read my mind and put me off in two jumps. I had a feeling he'd just started, so I sat down on a pitch stump, hoping he'd get his mind on another track.

I should have picked another stump. Yellow with pitch, it so impregnated my Levis with resin that, once I hit the saddle again, hard as he bucked there was no way he could get me off short of throwing me clear out of my underwear. It so demoralized the horse that by the time I hit the river and turned south I could have rolled a ciga-rette in one hand with him doing his worst.

Dismounting at the ranch, however, was a different matter. Every time I tried to get off, there was a ripping sound like adhesive tape being torn from the rib cage of an athlete. Sleepy panicked, smashed down a pole gate, and stampeded through the willows to the house spring.

It took the combined efforts of two smirking cowboys and a gallon of coal oil to dissolve me out of the saddle.

That night I spent my first night in the big stone house at Yamsi that would someday be my home. As I lay in my wooden bed, I thought back on those boys on the street corner in far-off Michigan. I had a feeling I had lucked out and triumphed over them all.

As I stretched out on a bed for the first time in weeks, the big diesel light plant ceased its monotonous rattling, the lights faded, and there was only the howling of coyotes on the pine-clad ridges to lull me to sleep.

COYOTES HOWLING AT NIGHT have a way of making one feel vulnerable. If they are up to mischief, and all of them sound as though they are, there is not a darn thing you can do about it until morning, when it is too late anyway. There are not many animals that announce their presence with such impudence. To the rancher, they are saying, "Sleep well, if you can!"

From that little bed at Yamsi I would hear one coyote chorus start down past the barns, and that would be picked up by another chorus up on Taylor Butte, answered by another chorus over toward the Marsh, and another over at Calimus. And so on, I suppose, all across the country.

When I think back upon that bed, I realize I saw darn little of it. At dawn, with dew on our boots, we staggered out to grease the machinery, harness the haying horses, mend the hay nets, and do the dozen other chores connected with putting up stacks of loose meadow hay. I did not appreciate it then, but I was learning a system of haying that would soon be a lost art.

Teams mowed the hay with six-foot mowers, raked it into windrows with dump or sulky rakes, then, pushing before them buck rakes with long, slender wooden teeth that jousted at the fallen grass, the horses combed the meadows, shoved windrows into piles, and pushed the piles toward the stack, sliding the hay onto nets, then backing out from under the load. The nets were fastened about the load; then the hay moved skyward on cables swinging from the arm of a tall derrick pole. Spotted into position over the stack, the loads were tripped and dumped, wherever the stack boss saw fit, to shape his growing edifice.

Pitchfork in blistered hands, I spent weeks on those stacks, waist deep in hay, itching but with no time to scratch, longing for the cool of evening. High overhead, the loads came swinging in. "Dump'er!" the stack boss would yell, and a hay hand would pull a trip rope which unlatched one end of the net and sent the hay crashing downward in a cloud of dust, chaff, and grasshoppers. Wallowing in with pitchforks, we worked the hay toward the edge of the stack, building a straight, even lip until the stack looked like a great smooth loaf of green bread, with the top properly thatched against fall rains.

Watching me work, the seasoned hands grinned. At every step I lurched drunkenly, trying to keep a load of hay on slippery pitchfork tines seemingly committed to losing it. Whenever I moved too slowly, the stack boss showed his evil grin, an unlit Bull Durham cigarette working in the corner of his mouth, and dumped the whole shebang down on my head.

I might have despaired, but one of the hands failed to show up from a Saturday night drunk, and I got his job driving a derrick team—two dapple greys whose function was to move forward and then back, pulling the net loads into the air and, when the hay had been dumped, lowering the nets gently back to earth to be reset for the next buck load. The horses knew the job better than I, and I walked behind them in a daze, hardly touching a line as I listened to sandhill cranes, blackbirds, and Wilson snipe calling me off to the river for a swim. The river I had traveled from Michigan to enjoy was only half a mile away, and sometimes the early morning fogs marked its course, but I was given no chance to see it.

One day as I despaired of ever learning another job, the boss handed me the reins of a buck rake team, and I soon learned the delight of working with good honest animals that knew their job intimately and seemed to take pride in doing it well. Amongst the draft animals on the ranch, the light Percheron buck rake teams were the classiest.

I had driven out for my first buck load when I discovered a new excitement. As my team pushed the wooden spears under a windrow to gather a load, a pair of coyotes rose from the stubble and followed along, not twenty feet behind me, catching the meadow mice that lay suddenly exposed. I marveled at their capacity to devour mice until I made a wide swing toward some windrows near the timber, and there I saw four little coyote pups playing outside a den. Every now and then one of the parents would slip in from the field to regurgitate a belly load of mice, then trot back to the field.

Sometimes the pair followed my team, sometimes one of the others, but I fancied they preferred mine. As they trotted along, tongues hanging out in the heat, I talked to them hoping for a tail wag of recognition, but I had only to pay them attention and they would drop behind suspiciously, often crossing a sea of windrows to follow another team. I soon learned that they followed best when I pretended I didn't see them.

During that summer I watched the pups grow until they too came out to follow the teams. But there was as much play in them as hunt. They spent a great deal of time wallowing in the windrows, scattering hay, wrestling each other down with gaping mouths, or trying to pounce on grasshoppers, which invariably escaped at the last moment to fly saucily away.

All too soon, the last buck load of hay was gathered and the stackyard gates closed to keep out livestock. The valley was dotted with stacks, bleaching tan with the sun, settling against winter, when they would be broken open and forked on wagons to feed hungry cattle in the snow. The hay crews took their summer money and were gone. I was sent back to the Marsh for gathering time, spending long, hard days in the saddle helping to cut out dry cows for market and doing the hundred other tasks that are the lot of the cowboy on a big spread.

The first chance I had to steal away, I rode over to the aspen thicket beside the railroad tracks and watched for The Railroader, but he did not appear. The next day I sat through three trains, but there was no coyote. To me, the Marsh had lost some of its charm.

When I confided his disappearance to Ash Morrow, he seemed unperturbed.

"Ah reckon yuh didn't see ground squirrels over there either, did yuh?" he pointed out.

"Come to think of it, I didn't!"

"They've already gone to sleep for the winter. Hell, they go to bed in August, those squirrels. It didn't take thet old coyote long to figure thet one out; I expect he's off now huntin' other game."

That fall, in September, a sudden, fierce blizzard roared down from the Cascades, catching me still in my summer jacket and turning the brown marsh grasses to white. Caught far from the bunkhouse, I sat hunched and miserable in my saddle, head bowed to the onslaught, trying to make myself small under my big hat. All landmarks were gone in the whiteout.

Suddenly, a host of stories I'd heard about the Marsh came flooding back. I realized that with the windchill factor, I could die out there from exposure; other cowboys had.

For a time the big flakes melted against my clothing, then, as the weather turned sharply colder, they began to freeze until I was covered with a rind of diamond ice. Off and on that morning, for the first eight miles or so, Sleepy had tried to buck, but there was no outlaw left in him now. He stood hunched up, eyes closed against the whipping ice particles, weathervaning his tail to meet each shift of the gale.

I was scared. The rest of the cowboys had gone off to town, leaving me to check the far fields for strays. Unless I found my way to the timbered shore, I would freeze out there before they came in late that night and found me gone.

The icy wind increased until I could hardly breathe and the horse had to lean back against the wind to keep its balance. All my sense of direction had faded away.

Above the wind I thought I heard a whine. With numbed fingers, I forced my eyelids open. Shadows in the whiteout! Three coyotes, eyes sealed with crusts of ice, came trotting by, nose to tail. Without being able to see, they were following a cow trail.

Maybe they knew something! Maybe the trail led in off the Marsh to shelter among the pine thickets along the shore. I had little choice but to follow. If they sensed my presence, they did not seem to care. Like animals driven by fire or flood, ancient fears or enmities seemed lost in an armistice of desperation.

Sometimes I lost them in the furious blasts and had only faint tracks to follow—tracks that one moment later would be swept away. There were times when I almost panicked, afraid that they were as confused as I, leading me farther and farther out into the mysterious bogs and sinkholes of the central Marsh. But with blind faith, I followed.

Soon their coats had iced so that they were white

against white, barely discernible against the accumulations of snow. I kept my head low, taking the brunt of the storm on my hat, peeking ahead under the brim every five or six steps, squinting until I saw the coyotes again, then traveling blind as I tried to blink ice from my lashes.

Then, suddenly, I fancied that the wind had lessened, saw the first trees, barely discernible, huddled together like Indian tipis covered with hides of white. Sleepy nickered and threw his last bit of reserve strength into trotting forward. As we moved into the shelter of the lodgepole pines, the wind seemed to die, and my wet cheeks burned hot. When next I thought to look for the coyotes they had vanished.

10

THERE WAS ENOUGH GOOD pitch in that worry
log to let its body age roughly at the same rate mine did.
One grows attached to horses and dogs and, as one's animal friends mature, grow old, and die, one has to put up
with several sadnesses in the course of a human lifetime.
The log, like me, has sagged a little and acquired a little
moss, but we both, like the one-hoss shay, may stick
around ninety years and a day.

The log had a nice, warm, friendly aura about it, tem-
pered perhaps by all the troubles I had brought it through
the years. White-footed mice and chipmunks raised their
broods in its cool, moist depths, squirrels hoarded pine
seeds in its caverns, and bore worms moved their slow,
patient way through the areas with less pitch.

Perhaps a couple of ley lines intersected where I sat, for
there was indeed something magical—an energy field
that took one over and soothed or inspired, whatever

one's need. I had only to sit down and relax, and I could drift away to whatever segment of the past I wanted to relive. I could conjure up Don Coyote and see him trotting along through the woods as though he were alive. He would look so real to me that I would have to watch the woods to see if chipmunks or robins and pine squirrels were being frightened by him before I could accept the fact that I saw only a specter.

Of course, my uncle also roamed those woods around the worry log, and I liked to think back on those times when I first came to know him.

Since my uncle preferred to run his organization from town, I saw him only occasionally in those days. Now and then, regardless of how busy I was, he would pick me up in his big green Chrysler and take me along to open gates, while my saddle horse stood patiently tied to a tree awaiting my return. On those excursions he talked volubly about the land and its history, but he kept his hearing aid turned down so that conversation was all but impossible. Years later, when I came to understand him better, I began to recognize his shyness as my own. He was as wary of me as I was of him.

That first summer, my busy life on the ranch, surrounded by nature, kept me from being lonely. I telephoned my parents occasionally, and they wrote often enough from home, but little by little their letters came to describe a way of life that was alien, even dull. My own letters seemed to me to burst with excitement: the horses I was breaking, the coyotes I saw hunting on the meadows, the men with whom I worked and whom I learned to respect.

I enjoyed the discipline of hard work and was considering devoting my life to being a cowboy. My blisters turned to callouses. Fall gathering meant sixty miles a day ahorseback at a gut ache of a trot, each man keeping five good saddle horses ridden to the bone. High in the bit-

terbrush ranges we rode the water holes, pushing the cattle along frost-burned, buckskin draws down to the fields
along the Williamson, where we weaned the calves, selected replacement heifers, and culled the herd. I did not
dare look beyond each day for fear something might
change.

I hadn't mentioned to anyone the reality of going to
school. The nip in the air and the signs of approaching
fall almost convinced me that everyone concerned with
my life had forgotten about it.

It was also typical of my uncle that he too avoided facing the issue. Then, one day, he arrived just as I was sad-
dling Sleepy, handed me enough changes of underwear,
shirts, and socks to stock a dry goods store, and shipped
me off to school in California. It would be years before I
knew the peace of living in one place again.

Looking back, I remember the joys of coming back
regularly to Yamsi, and the pain of leaving it. Vacations
at the ranch ended inevitably in heart-wrenching good-
byes as I turned my saddle horses out to pasture. During
those years I hit all the phases of growing up: voice
lowering, then dropping to a deep bass; becoming
clumsy gaited as a new draft colt; thinking I knew nothing, thinking I knew something, then, worse, thinking
no one else knew anything. In the end, at six-foot five, I
towered over those who had once towered over me.

At each homecoming, I saw with wistful sadness how
my uncle had aged, and during the war years, I watched
the western traditions change and the world I'd loved as a
young cowboy fast disappear.

The horses were getting gentle. Quarter horse blood
had infected and ruined the stock. Hell, there'd been a
day when if a man's horse didn't buck the first mile, he
knew the animal was either sick or getting too old to
ride. No longer did a cowboy have to stand way up by a
horse's ear to mount. The old, quiet hands with cattle
were gone; the new ones rode on nervous energy, work-

ing stock as though they were chasing a herd of stamped-ing longhorns on a movie set. The old-timers, who once lived out their days on ranches telling tales and keeping alive the essence of ranch life, had moved to town on Social Security and had been replaced by men of no particular tradition.

One year I watched The Railroader, stiff and lame now, gray in the face, tail witch-knotted, ratty as a porcupine's, but still hunting ground squirrels along that same stretch of railroad. He'd been there at the time of the last steam locomotive and seen it replaced by noisy, stinking diesels. The next year he was gone, and the land was lonely without him.

School gave me little more than a veneer. Underneath was the lonely boy from the forests of northern Michigan who loved all things wild. I felt a separateness from my race. When I looked at the land I saw the earth's energies as a visual, dynamic reality; with my cheek laid against a mountain, I could sense its discontent, feel its tormented bowels grumble, strain for relief; I could see roots of pine, aspen, mountain mahogany, and manzanita straining deep into pumice soil, quarreling and competing for remnants of last year's melted snow.

When a golden eagle sailed down off the ridges above Yamsi, I felt I had but to raise my arms to sail away with it. When I heard coyotes howling from the rimrocks, I felt that I had but to tip back my head and the same wild, spontaneous, uncontrollable ecstasy would come tumbling forth. As I walked my uncle's forest, I felt a mounting presence, until shadow people, from ages long past, stepped out of time lost and into reality—shades concerned about the land and the manner of change, mournful that we abuse the privilege of being.

11

MY UNCLE AND I were perhaps closer than either one of us would ever have admitted. If he was noticeably upset when I arrived home from the service after World War Two driving a huge army surplus truck, windows broken, canvas flapping in the wind, and the rear end occupied by a scrawny mare named Straight Edge that I'd acquired in Louisiana, he forgot his anger as the mare waxed fat on his lush meadows and each year heralded my return from college with a splendid foal.

When I married a girl who rode horses like a Comanche and loved the ranch as I did, the old man was about half proud, but he had a horror of population explosions, and every time my wife, Gerdi, became obviously pregnant, he burst out in wrath, forgot that by now I was foreman at Yamsi, and berated me like a child.

"Hell, you're just like your mother!" he'd roar. "Had a bunch of kids she couldn't manage and a husband that couldn't work. Dammit, you don't even have a job, and all those kids—!"

"I damn well don't have a job that pays me half what I'm worth!" I'd put in as my part of the ritual. I actually made a bit of money on his anger, for, sooner or later, I knew he'd start feeling guilty and slip me a much-needed check as his way of saying "I'm sorry."

Once he had accepted the new arrival, he took to looking at the child as part of the history of the valley. Since the ranch was surrounded by an Indian reservation and he had acquired the land from various Indian ownerships, he saw Dayton, Ginny, Marsha, John, and Taylor as the first white children ever raised there and made them feel special.

He dangled the ranch before me like a carrot to keep me from moving on. But his promises to sell me Yamsi had a way of evaporating, and as regularly as snowmelt in the spring, he would put the place up for sale. People came flocking in from far and wide to pay him court: the dreamers, the superrich, the merely curious, all professing to lust for that fantastic valley at the headwaters of the Williamson River.

Perhaps to turn the thumbscrews a little deeper, my uncle sent them to me, expecting me, as manager, to show the property and even brag about it, but knowing full well what anguish the act would bring.

I was so determined to own the ranch myself, I made sure the visitors saw through the beauties of the land to its economic realities. If they remained enthused, I invented mysterious diseases plaguing the cattle, drove the sightseers past the remains of every cow that had died in the last decade. Under the guise of giving them a shot of local color, I presented my neighbors as bootleggers, cattle rustlers, murderers, fiends, and degenerates. If a man arrived with a young, pretty wife whom he obviously adored, I dreamed up a neighbor who was a fantastically successful ladies' man. I learned to delve for each man's

fears and uncertainties and play to them. And I lied and
lied and lied.

Maybe I would have lost the race had it not been for
the meadow mouse invasion. One summer day, in 1958,
my uncle stood beside me, tall, gaunt, easing toward his
mid-seventies with grace, but sadly austere now, edged a
little closer to defeat by a frantic flood of meadow mice—
Microtus montanus—which tunneled a labyrinth through
his last green clover meadow. We were watching the
nearly complete desecration of Yamsi and what had once
been one of Oregon's most beautiful cattle ranches.

Before our eyes, the soil heaved and turned black, as
though churned and cultivated by a thousand invisible
plowshares. A chorus of shrill, angry squeakings filled
the air as mice competed for food and new territories,
then finding both scarce, turned cannibal and ate each
other.

Sinking ankle deep in the fresh tilth, the old man took
a few steps out into the ruin. Picking up a discarded
pitchfork, its shattered handle already gnawed to a whis-
per, he caught a dying rodent up with one rusty tine, and
silently, with distaste, pushed it toward me. Like most of
the mice around it, it had pink patches where the hide
had been bitten away. Extremities of limbs were gone,
leaving ambulatory stumps, but the mice, undeterred by
such handicaps, scrambled on.

I thought he was going to use the occasion to comment
on what the world was going to be like if Gerdi had any
more children, but he held his peace.

Even though my uncle was still owner and calling the
shots, I think he held me, as ranch manager, responsible
for this debacle. Nature seemed to have gone mad, and all
we could do now was watch and wonder where mankind
had erred. Klamath County, a lovely, rich agricultural
and forest area just east of Crater Lake National Park and
the Oregon Cascades, was caught up in a plague of mice

that, save for one reported epidemic in Russia, was the largest in world history.

If harried scientists referred to the four-inch-long animals as "Microtus," to us lay people they were meadow mice, tule rats, or voles. Whatever their name, they were a harsh reality upon the land and a people whose income depended upon agriculture.

Even in our isolated valley, twenty miles from our nearest neighbors, the mice population had burgeoned into nightmare proportions, with the old man and me caught up in its drama and powerless to do anything about it.

Rodent populations are normally cyclic. Various factors had combined to produce an extreme condition, with no one factor to blame. Heavy use of agricultural pesticides may have caused a dearth of fleas and other disease vectors which would normally be present in an ascending population. The great predator war of the past decade and the senseless killing of raptors had left the mice with few natural enemies. Rodent control practices such as poisoning with chemicals had actually kept the existing mouse populations healthy, preventing the build-up of natural diseases, which, more than predators, limit numbers.

All these factors contributed to the existence of an abnormally high base upon which an explosion could begin. A good feed year for mice sent populations climbing, and although a wet, cold, snowy winter might have sent numbers skidding back again, the ensuing winter was one of the mildest on record. Even the tops of mountains showed only patchy snow. Mice bred all winter, and six weeks after each litter was born, it too was ready to go into production.

Predator populations are effective only when they can keep the rodent population healthy and in line with food supplies and can reduce the highs and lows of normal cycles. Once an explosion is under way, control by larger forms of predators, such as coyotes, weasels, mink, eagles, hawks, owls, herons, cranes, and seagulls, is ineffective and gives way to more subtle forms of control such as disease, starvation, neurotic tendencies affecting production, and cannibalism.

Working just beneath the surface of croplands and pastures, the rodents tunneled frantically, searching for any stalk, seed, or root overlooked by the competition. Irrigation only concentrated the animals on ditch banks, which crumbled into a mass of tunnels, while the voles became ever more neurotic as their density increased and they rushed pell-mell for new territories only to find them taken by other voles.

Luckier than most, we had put up our winter hay early, and the cattle were out on the ranges until fall, when storms and cowboys would bring them straggling into the feed grounds. On the ranch proper there was not much we could do but hope that some natural calamity would happen to wipe out the invaders.

In other sections of the county the situation was equally disastrous, and the remedies tried were as varied as they were ineffectual. Driving to the local pound for a pick-up load of cats, a rancher would find only empty cages. Every available cat and kitten had long since found a home.

The county agricultural agents suddenly became the center of attention and led the fight with studies, news releases, sprays, and poisons. Sadly enough, the latter two killed not only the target species, but thousands of birds such as Steller's jays and mourning doves.

According to Martha Smith, then one of the owners of the Geary Ranch, one of the largest grass seed producers in the world, a spraying of toxiphene for mice left the ground littered with hundreds of dead geese.

Unfortunately for those trying to cope, control programs only seemed to keep the mouse population healthy and postpone the day when natural controls such as disease, cannibalism, starvation, or behavioral factors limiting breeding and production could take control. Winter was our great hope, but even the weather seemed to conspire against us: shirt-sleeve days and mild nights instead of the usual cold. The number of meadow mice continued to build, and whenever a local population exploded to its limit, the numbers burst outward to less crowded fields.

At Yamsi that winter we were feeding some six hundred head of Hereford cattle. The long stacks of baled hay in the sheds could have taken us comfortably through the winter, but now the piles began to sag like swaybacked horses. Mice moved in by the tens of thousands and so riddled the bales with tunnels that there was little seed or nutritive value left.

Break open a bale and inside was a cluster of mice blinking at daylight from a maze of chambers black with excrement. The cows rushed hungrily after the feed wagons, but only sniffed at the ruin of bales we fed over the side, then came trampling on, bawling their discontent.

Load a truck or wagon with hay, and mice decamped by the hundreds as the vehicle moved down the highway. There they were run over by automobiles, and the highways around stack yards became greasy with the mess. Cattle feeders tied twine around sleeves and pant cuffs. Women stayed at home, some even sleeping in bathtubs filled with water.

Ranchers went to their hay piles to find twine cut, bales loose, handles of hay hooks eaten away. Harnesses fell off draft horses in mid-load, saddles were devoured until only the cinch rings and wooden trees remained. Mice invaded automobiles, trucks, and tractor toolboxes, chewed wiring harnesses, and stole padding from seats for nests. One morning as I started the ranch pick-up to drive my uncle down into the fields, a mouse nest on the manifold burst into flames and destroyed the vehicle before we could summon help.

Invading houses, the voles chewed drapes from windows, blankets from beds, straws from brooms. They devoured the roots of shrubs and trees so that some ancient pines that had withstood many a tough winter storm fell upon each other like jackstraws. Fruit trees stood with branches white and bare as antlers of an elk. On the Geary Ranch, Martha Smith still recalls their helplessness as huge supplies of stored seed dwindled or were ruined. When some of the Geary workers took sick with raging fevers, rumors of tularemia exploded, and there was near panic.

The end of the epidemic came quickly, but not before the county had suffered millions of dollars damage to crops. Quietly and undramatically, the mice died out, rotting into the soil. "God's work, the whole disastrous epidemic," some said. But there were those who felt that we had somehow done it to ourselves, that the whole catastrophe was the price of our inability to understand natural checks and balances, inevitable reparation not only for our war on predators, but our refusal to ally ourselves with the forces that keep Nature's system working.

I stayed out of the arguments. There was work to do to heal the shattered ranch, trying to reseed the blackened fields and rebuild the riddled ditch banks so that they could carry water to thirsting lands.

Spring came and with it a time of rebirth. Down along the river, Canada geese nested by the hundreds, and sandhill cranes shouted out the limits of marshy fiefdoms. Normally I would have known most of their secrets, but there was little time to observe them. The cattle had to be branded and scattered on the ranges—such pasture cattle as the ravaged fields would support trucked in from California—and fences had to be repaired where mouse-damaged trees had toppled in the storms.

It was also a time of takeover, one generation from another. My uncle had lingered on past the years when most men retire, in part because he loved the land, in part, I suppose, because he was waiting for one good year to happen. The mouse epidemic was a hard blow to his hopes.

12

AFTER THE MICE HAD reduced the fields to rubble, and after a would-be purchaser, having talked to me, had passed my uncle on the way to town and shaken his fist at him out the car window, the old man gave in and sold me the ranch.

"Hell, you'll starve to death!" he shot at me as I rushed out of the house, signed agreement clutched to my chest, to tell my world—a wife, five children, and two Indian cowboys—that I had bought the property.

My wife's eyes shone with her happiness, and the children danced about, relieved that they would not have to finish out their childhood elsewhere.

I was happy sad. A faint mistrust of my own enthusiasm plagued me. However beautiful the land and the lovely river that flowed through it, there were basic problems of production and management we would have to solve if we were to survive. Reality washed over me, leaving me sputtering with doubts. The idea wasn't just fun anymore.

The land I walked on felt different now that I owned it. Drifting quietly away from my family, I moved off into the forest and sat on the old worry log, gazing off down the valley. The dying sun left a pinkish memory in the air. A few red-winged blackbirds sat on the top wire of a Page wire fence, then departed roostward, conversing with each other. Just overhead, a nighthawk startled me with its booming. I followed it with my gaze as it glided and darted its haphazard, open-mouthed way, hawking invisible insects above the silent pines. Suddenly, I had a strange feeling of being watched, and I turned slowly to see a large coyote, standing with both forefeet on a log, peering over at me.

Once he had made eye contact, he woofed nervously and ducked back behind the log. When I saw him again, he was using the log for cover and was trotting down toward the meadow, casting back an occasional glance to make sure I intended no harm. Pausing at the edge of the timber, he sniffed the ground, then proceeded to make the earth fly with his front feet as he dug furiously, tearing up the meadow as though it were his turf, not mine. I realized quite suddenly that I wasn't the only owner of that ranch.

Looking halfway down the valley, I could see the shine of water, as the river coursed over thirsty fields, and the tumbled-down old stack yards from the old days of loose hay. In the silence I heard my uncle's car cough and purr, then head off up the road toward town. I should have been there when he left to thank him and say goodbye. I knew him well enough to know that he might never come to the ranch again.

Hoping to head him off, I ran up the hill toward the road, but I was too late. Through the dusk I saw him pass; as I reached the road out of breath, I saw the glowing red of his taillights as he braked for the first bend and disappeared. For all those years and years he had owned the ranch, and now there was someone else in command—a kid from Michigan he hadn't really invited West.

It was almost dark when I returned to my worry log to think and to plan. With my shoestring economics, there wasn't much margin for error. I had to understand that land as I had never understood it before. It was time to recognize that, like so many other ranchers, I'd been fighting my land too much.

Take the short-tailed ground squirrels that plagued our pastures, for instance. Each and every summer we walked miles, putting a teaspoon of poison oat groats in each hole. According to studies at Oregon State University, twelve ground squirrels per acre, on bluegrass pastures, reduced the forage a whopping thirty-seven percent, enough to make economic ruin of a hay crop. Yet whatever our attempts to murder the last ground squirrel on earth, and although their numbers would decline by the end of the summer, the following year the animals would recover, often in unprecedented numbers. Another epidemic such as that of the meadow mice would ruin me. I wondered what would happen if we flat out quit poisoning and let the natural systems take over.

In the past we had taken a lot of advice from agricultural extension people, spraying all manner of hard insecticides on horn flies, face flies, horse flies, mosquitoes, lice, ticks, grubs, and grasshoppers. Such ventures seemed to bring only temporary relief. In the end, all we seemed to do was keep the pest population healthy.

A big yellow moon soared up behind me through the trees, revealing the landscape once more. Down on the meadow, the coyote was still digging. Against the green I could see a great yellow spill of corn pumice where the animal had dug down through the thin black topsoil, a mound as fresh and clean as though it had been blown from Mount Mazama yesterday. I felt betrayed. What

had I ever done to coyotes that they would tear up my
fields like that? There had once been something so neat
and ordered about that pasture. Just last year it had been
fence-to-fence unbroken greenery. Between the mice and
the coyote, it would take three years to heal from such
unthinking vandalism.

I felt a sense of relief when the coyote caught whatever
it was it had been seeking and killed it after a brief scramble. And now what was it doing? Digging another hole
to administer last rites in? Yesterday I would have smiled.
It had been someone else's ranch. Now things were different.

"Hey, you!" I shouted, coming off my log. The coyote
retreated into the gloom and stood calmly assessing my
chances of catching up with him, then, blithely, began
digging a new hole.

I turned away and headed back toward the ranch
house. On the morrow I had things to do. I'd been aching to disk and plant a piece of hummock land with
Reed's canary grass, which the county agent swore could
end our chronic winter shortages of hay. That night I
dreamed of natural disasters: frosts blackening the valley
in a return to the Ice Age; heavy winter snows suffocat-
ing the cattle with a merciless white shroud; drought,
grasshoppers, ground squirrels, and meadow mice turning the valley to dust—a silent valley where no bird sang,
and where no coyote dug up my meadows or lifted his
mournful voice to the wind.

13

I WAS TO SEE that big hole-digging coyote around my worry log again, of course, for that was the animal that wormed himself into my affections and became Don Coyote. It was during my first summer of owning Yamsi that my affair with the wily beast bloomed and during my first autumn that tragedy put an end to our friendship. My feelings on seeing him gunned down by the hunter went far beyond anger, sorrow, and a sense of personal loss. With Don Coyote, I felt I had been on the track of something elusive—an understanding of the land that fate had given me to own and manage. With the end of our relationship, that insight moved tauntingly just out of reach.

During those next hard days after he had been shot by the trespasser, and with only my sorrow to distract me, I buried myself in my work, spending long hours with my wife and children getting the cattle ready for winter and hauling wood for the big, hungry ranch house furnace, which made slaves of us all.

Down along the Williamson, the ducks and geese made one final circle about the ranch and moved south beyond the grip of the cold. Soon the first snows of winter would drift down off the shoulders of Yamsay Mountain, covering the ranch with a white blanket until spring.

One morning in December I stood looking out the front window of the ranch house at the leaden skies. I could feel in my bones that the weather was about to do something spectacular. "Guess I'd better try to move Big Alice up around the house," I said to John. "Might need it one of these days to plow snow."

John's eyes twinkled. "How about the parts for—what was it you called it—the dinkerator chuck?"

"It will have to run without it," I replied, turning my face away to hide a smile. "Better come with me. All the months that old tractor has been sitting idle, I may need help in getting it started."

As we drove to the tractor a light snow was falling, giving the bleached land a new look. Out of the frosted pick-up windows I kept seeing animal tracks, but the fresh flakes made dimples of them. They could have been made by a deer or even an errant yearling late in coming in off the range.

In the distance, Big Alice looked cold and uninviting. Thanks to the wind, I could expect a blanket of snow on the tractor seat where it had drifted under the canopy. I wished I'd been smarter and brought along a windbreaker and waterproof pants.

The wind came howling down off the mountain, driving the snow before it across the treeless meadows until the ruts of the road filled brim full and John had to guess at the trail.

We had nearly gained the tractor when, suddenly, John jammed on the brakes. "Hey!" he said. "What's that over by Big Alice? I saw something move!"

"The brakes work," I grumbled, pulling my head out of the glove compartment, "now try the horn."

For a moment a blinding fury of snow blew across the flats, obscuring everything.

"Really, Dad! I saw something!"

I strained to see out. Nothing. White on white. Then suddenly the wind seemed to suck in its breath for a new blow, and in the lull, I saw a dark, moving form against the snow. Scrambling awkwardly to gain the shelter of the tractor was a coyote. Its tail had been cropped off at the base, and it was stumbling along on three legs.

"It's the Don!" I yelped. "I can't believe my eyes. I left him for dead!"

Out of habit, I patted my coat pocket for a piece of bologna or a sandwich, then realized I hadn't packed any since Don Coyote's tragedy. He had to be half starved, and the weather for hunting wasn't getting any better. I wanted to bring him some carrion, but there wasn't a dead animal on the place.

By now the defroster could only partially handle the falling snow, and our visibility was restricted to two grapefruit-sized patches on the windshield. Right now, there wasn't a thing I could do for the Don.

"Let's head for the barn, Sir John," I said. I could have used the big tractor to plow snow off the feed grounds, but Don Coyote needed it worse than we did. He was back in my life, and now all I had to do was figure out how to keep him alive.

14

AS EXCITED AS I was to find that the Don was still alive, I was haunted by my own part in his torments and steeled myself against attempting to renew our old relationship. A gift of carrion maybe, but no more bologna or sandwiches. No more meddling in his affairs. Period!

What hurt me deeply was the thought that the Don would associate me with his agony and shun me forever. I gave strict orders that no one was to go near the tractor, but the next day I broke that edict myself. As I rode down the valley on old Straight Edge to look for strays, I could not resist trying to find his tracks where he had crossed the snow field ahead of us the afternoon before.

However, the storm had done its job well. Once or twice I thought I saw his fleeting shadow in the forest, but every small pine was a snow cave of drooping, encrusted branches, and always the image vanished before my eyes could focus.

Once, feeling disconsolate and alone, I tried my best to howl, but the hills flung back my voice as though saying, "Let him be!" There was no answer from bird or coyote. Since the poisoning, the wildlife had been slow to move back, and I heard only the forlorn dirges the wind sings as it explores the upper reaches of pines. I longed for the merry pipings of a convocation of chickadees and nuthatches, the rattle of magpies, or the cork-popping of a passing raven. If, from his hidden retreat, Don Coyote heard me at all, I hoped it would give him comfort.

The kids had made bets amongst themselves as to just how long I could stay away from meddling.

"Maybe his wounds are infected, Dad," Dayton suggested, grinning at the others.

"What if he's lying down a hole somewhere and a little medication would help?" Taylor offered, always the veterinarian.

"How can you sit here like this? Go to him!" Marsha said dramatically, pointing to the door.

"Here," my wife said, thrusting a whole ring of bologna into my hand. "At least let him know you care!"

"If his wounds are infected," I said, brightening a little, "then a spot of medication hidden in the bologna wouldn't exactly be meddling in his life, would it?"

From a pine tree outside the house I stripped a few handfuls of pine needles; from a drawer beside the kitchen sink I borrowed rubber gloves, then doctored the bologna with sulfa pills two inches long. Boiling the pine needles in a pot on the stove, I dipped the meat into the swirling cauldron to destroy human scent.

"What on earth are you up to now?" my wife asked, gagging as she entered the kitchen. "It smells like a pack rat den in here."

I left my gift of love at the tractor, and next morning it was gone. Perhaps it had been stolen by a passing raven, but at least I felt better. I could only hope for the best, that in some secluded corner of his burrow, Don Coyote was fighting attacking microbes and winning. Now all I could do was wait.

In late January I thought I saw him again. Dismounting quickly from Straight Edge, I took my binoculars and moved to a vantage point on a timbered hillside where the mountain sun had melted the snow from a patch of pine needles and I could scan the area in some comfort.

The mare looked at me with curiosity, but turned herself broadside to the spring sun and dozed, her lower lip hanging in contentment and one hind leg akimbo. Old age was creeping up on her, and the fat was melting away, leaving her inner structure of bones too obvious to help her looks. I was reminded of when I had seen her first, standing half starved in a tumbledown corral in Louisiana, so many years before.

For some minutes I lay watching the forest, glassing the spot with binoculars where first I had seen movement, but now the woods looked deserted. A pine squirrel dropped a pine cone on my head, then fled to a higher branch, chuckling at its audacity. It began to scold, announcing my presence to the whole forest. I felt that every wild eye had to be upon me.

Suddenly, in a clearing where some great wind of yesteryear had mowed down old growth pine like jackstraws, I saw movement. A coyote leaped a log, dashed forward, caught a mouse, tossed it into the air, snapped it up again, and gulped it down. It had all four legs and a tail, and as it squatted and melted a yellow daffodil in the snow, I knew it to be a female or at least a pup who had not yet learned to lift its leg.

Moments later the coyote caught a pocket gopher, swallowed it without fanfare, then trotted toward the tractor. As it approached, it became nervous, weaving cautiously back and forth, watching the tractor carefully, ready to spook and run. Then its ribs began to heave with rhythmic convulsion, and it regurgitated a pile of food on the ground. Cocking its head, it listened as

though expecting Don Coyote to come out and accept
the gift. At length it seemed to tire of waiting, and without a backward glance, trotted off across the meadow toward the forest.

For half an hour nothing happened. The cold seeped
through my heavy coat, chilling me. Hugging my knees
against my chest, I tried to tough it out, fighting hard to
convince myself that the Don still lived.

Another half hour of misery, then a raven, passing
high and headed toward Yamsay Mountain, slowed, circled a few times, and dropped to gobble up the offering. I
cursed under my breath, trying unsuccessfully to drive it
away with malicious thoughts. For some minutes the raven strutted back and forth on the tractor roof as though
trying to fathom how such a spendid repast came so
easily into his day. Soon the distant croaking of another
raven called it away. Moments later I saw them both,
diving and playing along the cliffs of a cumulus cloud
high over the mountain, then they were lost to view.

I had almost decided that the Don had died in his den
and the female's efforts to feed him were in vain, when
suddenly he peered out from behind the bulldozer blade
and limped painfully into full view. He stood very still,
blinking at the shaft of spring sunshine that burned
through the clouds and bathed his immediate territory
with a spotlight from heaven. When he moved, it was
stiffly, as though he were in pain. Hobbling to where the
female had deposited the gophers, he seemed to read the
whole story in scents.

His hurts seemed to overwhelm him, and he lay down
on his side, licking his stump of a leg gingerly with his
tongue. His tattered coat showed his ill health, and I lay
frustrated at not being able to help him.

I was so busy watching the Don that I failed to notice
the little female trotting across the meadow. She approached him meekly—wagging her tail, wriggling her
body, showing her teeth in a smile—and crawled toward

him on her belly in subservience. Don Coyote rose, lowered his head, mouth gaping wide, hair standing straight along the ridgepole of his back. Cheeks puffed in a snarl, he looked twice as large as life. Whenever she tried to approach, he kept her at a distance with short rushes, as though he were afraid the female would bump his wounds.

Approaching as close as she dared, she once more retched up some food. He snarled his thanks, drove her back from the pile, and gobbled it voraciously as though starved. She sat some distance away looking pleased, her tail wagging with the slow, measured beat of a windshield wiper on low. Soon she rose, yawned, and trotted away again, nosing at grass clumps showing through the snow on the hillside, either hunting mice or simply reading scent posts, as though the grass clumps were mailboxes and she was interested in who lived there.

Four times in the next hour she returned with prey. As far as I could tell, the female was less than two years old and had never had a litter. I marveled at how strong the family instinct was amongst coyotes, where even young of previous litters sometimes help an adult pair raise pups. There was no pair bond here, but still the young female was taking care of an injured member of her species.

Always the procedure was the same, however. Don Coyote stole the offering away from her with a display of male chauvinist bad grace, as though to accept graciously would have been to accept charity.

He kept his home to himself. When she circled the tractor, sniffing as though to locate his den, he snarled and drove her back, his stub jerking half circles in the air as he moved.

She had been gone about an hour when the sun disappeared behind the clouds with an air of finality that suggested it would not be back that day. Don Coyote limped back to the tractor and vanished beneath it; I crept away, relieved at last of some of my guilt.

As though to punish me for snooping, the weather turned wintry with a vengeance. However snugly I dressed, I could only endure so much time lying in a snowbank. The cold seeped through my chaps, numbed my hands through woolen gloves, and turned my feet into chunks of stove wood. I built myself a little hut of snow with an observation window of glass, but although this lessened the chill factor from the wind, the hut grew dank with my condensing breath, and my few comfortable moments were spent outside the house not in it, when the mountain sun took pity on the land and sent down a beam of warmth.

Despite his atrocious behavior, Don Coyote fared well from the faithful ministrations of his friend. Day after day she came to feed him, and such loving care began to tell on his condition. His first trips away from the tractor to defecate and mark his territory were short, painful, and aimless. He could not seem to fathom that the leg was no longer there and that he did not need to cause himself unnecessary pain by jerking the stump as he walked.

Soon, however, he acquired the knack of traveling with one hind foot doing the work of two and was able to cross the meadow in deep snow. At last, one day as I rode by I saw the pair out hunting together, and it was the Don himself who trapped a pocket gopher in a snow tunnel, dug it out with his forepaws, and made the kill. The female sat watching intently nearby, her tail wagging, as Don Coyote ate the whole prize without sharing a morsel.

Two weeks passed, and still the cold weather punished the land. As though the cold invigorated him, the Don traveled farther and farther away from his den, with the female traveling at his side. One of their favorite hunting areas seemed to be a heavily logged area where snow covered a plantation of young ponderosa pines. The plantation had been written off as a near failure due to an invasion of pocket gophers dining on the roots of young trees.

Armed with a ground sheet and an arctic sleeping bag, I set up a blind in a thicket where I could scan a large area of the hillside without revealing my presence. I was eager to discover just how a coyote could harvest a pocket gopher working underneath the surface of the ground when the soil in turn was covered with two feet of crusted snow.

I spent two cold mornings on the hillside to no avail. On the third day my routine coincided with that of the coyotes. I located them traveling abreast of each other some twenty yards apart, relaxed yet alert, pausing from time to time to sniff the snow and listen. When one showed interest in the snowfield, the other would leave his own search and cross over to his partner. At times both would freeze, noses extended, pointing like setters, studies in concentration.

Often one would circle ahead and work back toward the other, as though trying to trap the prey in between. Often both would pounce in unison, scramble wildly, and make a kill just beneath the surface.

It didn't make sense. Why, for instance, were the pocket gophers leaving the safety of their earthen tunnels to hazard the snowpack where there were no roots to devour and where there was much more chance of being eaten?

One day as I was sitting in my blind, watching as the pair hunted a slope some three hundred yards distant, I was startled by an angry squeak in the snow beside me. The crust groaned as an animal tunneled past me, just inches from my knee. With one mittened hand I batted the animal out into the open. Out tumbled a large pocket gopher. Right on his tail was a little white weasel, no bigger around than my thumb. The gopher was twice his size, but without a glance at me, the weasel darted in for the kill.

The gopher clawed, squeaked, scrambled, and thrashed, but the weasel clung to his neck, and soon the animal ceased to struggle.

For a moment the weasel stood upright as a stake, eyeing me curiously, then, swiftly, it settled to work and dragged its prey to a hollow log. From the depths, I could hear sucking sounds as the weasel dined, and now and again he would pop out of one hole or another to stand and stare at me curiously.

I might have dismissed the event as being unrelated to the coyotes, but suddenly it hit me just why the coyotes were in the area. The gophers were attempting to escape the weasels by tunneling up into the soft snow just under the crust, where travel was easier. By pursuing the gophers, the weasels did much of the work, while the coyotes made the most of the opportunity and harvested the crop without giving the weasels a share.

15

SPRING BLEW IN ON a chinook wind out of California and caught me with some of my winter jobs still undone. On a ranch you take the worst emergencies first, and much as you plan developments, most of your time is taken up just undoing the disasters of the day; by nightfall you are ready to fall into bed with exhaustion but haven't accomplished much of what you set out as a goal.

One thing I had gained over the winter was a better understanding of how coyotes managed to hunt in the snow and survive, and I had come to appreciate their adaptability.

Before Don Coyote was shot, he had been a good study subject, for he went about his life basically as his parents had before him, and since he treated me as though I were invisible, I learned a great deal about coyote behavior just by watching him.

But now that he had suffered at the hands of humans, he was much harder to observe going about his business, and his behavior as a crippled coyote could not be considered normal. For study purposes, what I needed to do was get a pair of coyotes, breed them in captivity, and raise the pups under varying degrees of wildness by socializing some and letting the others of the litter go wild.

I had hoped that Don Coyote, by mating with the little female and producing a litter, would make it easy for me to borrow a few pups, but she was apparently too young or not in good condition, for she failed to come into estrus, and even during those weeks of coyote madness in February seemed less a mate to Don Coyote than a hunting companion and friend.

If on occasion he read her body signs with his nose and followed this ritual with territorial wetting of any scent posts that happened to be handy, he was not visibly interested.

During the winter I had corresponded with a few acquaintances who kept coyotes in captivity, in the hopes I could acquire a mated pair, but so far my efforts had been fruitless. Burdened as I was with ranch projects that needed doing, I took the time to write a few inquiries, but that was all.

An immediate problem was the water situation in the Calimus field. The only source of summer stock water was a deep well, and that was only as reliable as the gasoline engine that ran the pump. I had intended to overhaul the monstrous thing during the winter when work was slack, but somehow Don Coyote had kept me occupied. But it needed doing, and I tried to make time.

My plan was to snowshoe down to the pump house and carry the few vital parts that needed repair out on a toboggan, leaving any heavy work for later, but that proved harder than I imagined. The melting snow water had gutted the bridge, separating me from my work. Looking for a place where I could cross the torrent on rocks, I followed the snow-water river through gloomy thickets without finding a crossing and eventually found myself at the south end of the property, where the water poured down through a rocky canyon.

As I sloshed out across a small meadow, packing my snowshoes and careful not to go in over my boots, a pair of coyotes fled for safety, splashing out of the shallows where they had been wading. Their flight, in turn, precipitated a band of wild horses which wheeled about to determine what there was to spook at, then charged off over the hill with great poppings of brush and thumping of hooves. I glimpsed a sorrel stallion, scarred by battle, driving some blue velvet mares, then, swiftly, they were gone.

Amongst the ponderosa pines along the slopes, a few weathered lava rimrocks showed. There was an air of mystery about the place, and I had the strange feeling that ancient Indians watched me from along the rims. I climbed a fortress of gray rocks on the west rim and sat quietly for several minutes listening, watching, wishing the feeling of dread would go away. From the stream along the canyon bottom, the thin silvery music of water ouzels rose above the thunder of frothy golden snow water.

On the ground at my feet, I saw a curious grooved stone and picked it up to examine it. It was an Indian fishnet weight! How many thousands of years had it lain here undiscovered? And why was it here, so far from any existing lakes and permanent streams?

A canyon wren flew into a pile of rocks beneath me, dislodged a tiny slide of talus, then burst into its eerie song, as though laughing at how it had made me nearly jump out of my skin. At the point where the wren vanished into the rubble, I noticed a strata of diatomaceous earth that ran a few short yards, then disappeared under pumice sand. Instantly curious, I scrambled down the steep slope and knelt to examine my find. The rock wren fluttered out from under some talus, eyed me a moment as it swallowed a spider, then flew off scolding.

The net weight wasn't out of place after all. There had once been a lake here. Closing my eyes, I caught an instant picture of the land as it must have been thousands of years before. Towering high to the northwest, the gigantic Mount Mazama, as yet unshattered, dominated the skyline; here was the source of thunder, gathering clouds, dispensing wind and weather, and sending roaring avalanches of ice into the valleys below. Around me, a vast rolling desert of ancient lava was set in a matrix of red clay, its tortured junipers gambling against the next drenching of ash, the next molten flow.

Most important to me was that those lines, half buried in detritus, revealed the shoreline of an ancient lake. In my mind's eye the lake stretched out before me, mirroring the grandeur of the mountain. Along its shores were Indian villages set snug against lava outcroppings to gather and store heat from the sun. On the mirror surface of the lake, Indians, in canoes made of hollowed out logs, gathered wocus or water lilly seed and flung their nets weighted with grooved stones, casting for fish.

What had happened to the lake and its culture? I could only guess. Perhaps when Mount Mazama exploded into the ruin it is today, leaving only the huge caldera that now is Crater Lake, the rushing torrents of melting snows cut through a ridge, forming the Sprague River Canyon and drained the lake forever. For a time, the area that was to become the ranch property had lain as a great red flat, then subsequent explosions had covered the alluvial clays with a four-foot blanket of pumice sand. Apparently my basin had once been the bay of a large inland lake.

Ever since my boyhood in northern Michigan, I had been seeking to retrieve the lake culture into which I had been born and my roving eye had been looking for good dam sites throughout the arid West. I'd looked from train

windows, automobiles, airplanes, and the backs of
horses, unconsciously perhaps, but always trying to find
a place to build a lake. Now chance had shown me a potential site right in my own backyard.

I was about to search the site further, when I stepped
over a fallen log to find myself face-to-face with a big
black bear. His pelage was tattered from winter and his
disposition, after hibernation, grouchy. He stood upright
to peer at me. I remembered the Indian legends about
grizzlies in that valley and fled as though he were right on
my tail.

Back at the pump house, a pack rat scurried for the
rafters as I entered, and in the deserted bluebird nest in
the toolbox on the wall some white-footed mice peered
out from a pile of tow. Never an enthusiastic mechanic, I
worked slowly, thinking, as I loosened each rust-frozen
nut, that if I managed to get together enough money to
build a dam and catch the snowmelt, that abominable old
engine and its pump might one day be obsolete.

By the time I had reached the ranch and dumped as-
sorted pump parts needing welding in the blacksmith
shop, my excitement at the thought of building a lake
had reached fever pitch. Why not a lake? It was the one
thing out of my boyhood to make my land in Oregon
perfect. A more rational man would have settled for a
small livestock pond, damming up just enough water to
take care of the needs of his cattle. But I thought of the
scourge of raging snow water that for years had been
gutting channels in the basin, and vowed that someday
the water would be dammed and that some July, when
there ordinarily wasn't enough water left for a chickadee
to drink, wildlife would come from miles around to
drink from and wonder at my lake.

I longed to share my excitement with my family, but
when my wife met me at the front gate, the look on her
face made me keep any new ideas I might have to myself.
When she was upset with me, it was best to be silent.

"Who are the Kiracoffs?" she asked abruptly.

"Old friends. They run a private zoo in Pennsylvania. Are they arriving?"

"No, but they telephoned. They have a pair of homesick western coyotes they want to send you. Naturally, I turned them down."

"You turned them down?" I forgot the lake at Calimus and rushed for the telephone. Already my mind was racing: A beautiful pen by the house spring, with a neat kennel in which the bitch could whelp and a one-way glass so I could watch. I could hand-feed some of the pups, raise them semi-tame, and do all sorts of behavioral studies. At a time like this, a practical, sensible wife like Gerdi could certainly be a pain.

The coyotes were already on their way west by air freight when she came up with the latest of a series of good reasons why we didn't need those coyotes.

"Permits!" she said triumphantly. "I checked with the state game commission, and no way do they grant permission to keep coyotes in captivity. There's a state law against it. Too many enterprising people were raising litters and collecting bounties for their pups."

"No problem," I said. "Not only is their director a warm, personal friend, but he's also a very reasonable person. He wouldn't dream of turning me down." I turned my back on her and picked up the telephone.

"Hey, old buddy," I said when his secretary put me through.

He listened patiently enough until I mentioned getting a permit to keep coyotes, then he lost his cool. "No way," he shouted, loud enough for my wife to hear. There's a state law. What did you say your name was?"

I hung up the phone without answering. The coyotes were on their way! They would arrive the next day, and now there was every chance they would be confiscated at the airport and destroyed.

That night I tossed and turned, my dreams filled with

wild scenes. I had shot my way out of the air freight office, and now the police were hunting me with dogs; behind those barking hordes came a mob of angry sheepmen. With the pair of coyotes by my side, I slipped away into the back country, determined that if they should die, I would die with them. I had almost reached the safety of the lava brakes and tablelands, where one man could have held off an army, when an armed band, led by former friends who knew my ways, cut me off.

"Halt or we'll shoot, coyote lover!" a harsh voice commanded, echoing and reechoing from lichen-encrusted rims. I'd have known that voice anywhere! It belonged to my wife. "Surrender! Give up those damn coyotes!"

"Never! Go ahead and shoot."

"Wake up and scratch my back," my wife murmured, snuggling close.

The plane had already arrived at the Klamath Falls Airport when I drove up. At the door of the terminal, four husky men lounged with what seemed to be a studied disregard of my arrival. I had a feeling that they were wardens and I was walking right into a trap. I was ready to pay a fine for my activities, but I hated the thought that the coyotes might be put to death.

Behind the clerk, I could see a steel barrel stout enough to hold a tiger, its open end welded shut with heavy steel bars. No neat little aluminum dog kennel that! It was the sort of cage that would attract the curious. Behind the mesh I could see two sets of yellow eyes.

The four men slipped through the door behind me, then a fifth, a tough-looking roustabout type in faded coveralls, joined them.

"Relax," I thought. "They don't get shipments of coyotes in here every day. Maybe they work around here and are just curious to see coyotes up close."

I moved up to the desk. "You got a couple of little dogs for me?" I asked, sounding as though I'd been hit in the Adam's apple with a baseball.

The coyotes were so taken with the sound of my voice, they started to howl.

I didn't know what to do. It sounded as though a whole litter of dalmation pups had just been backed over by a fire truck.

Hoping to soothe the coyotes, I sat down beside the barrel and howled with them, but the animals only howled louder. Magnified by the steel barrel, the sound could have shattered glass.

"Whatcha going to do with them coyotes?" the clerk asked.

"Coyotes? Oh, those. Research. We're going to cross them with Hungarian parslapps. Breed a dog that looks like a coyote, howls like a coyote, but is harmless. We'll use 'em for decoys. Call in every coyote for miles, then: Bang! Bang! Bang! I sprayed the wall with imaginary bullets.

"Sheepmen would like that," a man said, impressed. "Got a cousin raises sheep over in Lake County. Can't wait to tell him."

"I'll save him a pup from the very first litter," I said. "Best way is to raise these dogs the way they do guard dogs—right with the sheep."

It is said that coyotes never foul their dens. It had been a long and frightening trip, yet the coyotes had contained their eastern waters with nary a mishap. But now either my howling brought out a sense of territory on the part of the male or just plain frightened him into incontinance. A large yellow puddle drained over the side of the table, splashed down upon the concrete floor, and began to travel. Ammonia fumes wafted through the air. Even the coyotes were silent as though with embarrassment.

The agent looked angry. Walking on his heels, he grabbed a push broom and tried to keep the advancing flood from reaching a pile of baggage on the floor. "Get those damn critters out of here!" he snapped.

Obligingly, the four spectators loaded the dripping cage into my station wagon. I slammed the door. I was getting away! The men weren't game wardens, just spectators, waiting, perhaps, for packages of their own. I handed them a fistful of money for their trouble and sped off down the road.

The heavy, junipery, pack-ratty odor of coyote urine was overpowering. My eyes watered until the onrushing traffic became blurred and the windows seemed fogged with steam. Along the highway I stopped at a gas station and bought an automobile deodorant guaranteed to make your car smell like a citrus grove, but the combination of the two odors was ghastly, and the blowers only contrived to concentrate the fumes in my face.

A cold spring fog off Klamath Lake turned the countryside into a deathscape. Roaring up beside me, a state policeman peered over at me, then cut off my escape. Mercifully, he only toyed with me, then sped on up the line. He settled for a yellow Volkswagon, never guessing that he had just passed up one of the great busts of his career. A grateful sheep industry would have deluged him with spring lamb.

In a field beside the highway, a few dainty lambs skipped and gamboled in newborn joy. They had about as much starch in them as a spaniel's ears. For hundreds of years Man had been breeding them weak and defenseless to keep them manageable. Without predators, they would continue to go downhill. Maybe the time was right to select for stronger lambs and for ewes sensible about hiding their lambs in the brush or otherwise capable of defending them against predators. Worry about how to catch them and take the wool off later.

The sixty miles seemed like a thousand. The big log gate at the entrance to the ranch was a welcome sight as I dashed through with the coyotes. It was my world now,

and I doubted whether the law would brave the barking
ranch dogs to spy on me.

The coyotes seemed to sense that we were off the
pavement and near their journey's end. They began to
whine and fidget in their barrel. As I drove up in front
of the new coyote run, all the tensions seemed to flow
from me, and I slumped against the wheel of the car in
relief.

16

"OOOOOOF!" SAID DAYTON AS he opened the car door to get a look at the coyotes. Eyes watering, he sat down on a nearby stump. "Keep those windows closed tight, Dad," he advised, "or the government will get you for polluting the atmosphere. I've never smelled anything that potent in my life."

I staggered out of the station wagon, sucking in my first full breath in many a mile. "Help me with this shipping crate," I commanded. "Those poor animals must be about asphyxiated. What they need now is a few good lungfuls of Oregon air."

Together we unloaded the barrel, placed it in the middle of their run, broke the welds on the door, opened it, and beat a hasty retreat so that the animals could leave their steel cage whenever they felt secure.

Cautiously, the big male stuck his long nose out the door, then the rest of his head and one paw. For a moment he surveyed the scene, then leaped boldly out. He made a few loping circles to determine the confines of his pen, then sat down in the middle of the run to scratch.

This seemed to indicate to the female that all was well, for she too emerged, although more timidly, ready to duck back into the barrel at the first sign of danger. Glancing at Dayton and me, she modestly hid her head behind a tree and urinated.

High in the pine tree, a squirrel scolded indignantly and bombed them with a half-shredded pine cone. As though already aware of the limits of safety, it followed the cone downward but became frustrated when the cone slipped through the wire mesh and struck the male coyote on the nose. The coyote ignored the insult and pretended to be looking off across the meadow, but when the squirrel descended to the wire, the big male leaped high and sent the trespasser racing giddily back up the tree.

Then, as though they had been born and raised in this meadow and had no particular desire to explore it further, the pair stretched out in the dappled sunshine and stared off down the valley seemingly lost in daydreams. Other than to glance at us when we moved, they paid us no more attention than if we had been part of a Sunday afternoon crowd at their former home in the zoo.

I named the pair Jack and Viola, after the Kiracoffs, who had sent them to my world, then left them alone for a time to adjust.

The spacious run gave them a great view. Just down the slope, the house spring bubbled and boiled from a witch's cauldron in the lava rock and moved off placidly through green, willow-bordered meadows.

Along the quiet pools there was a constant play of waterfowl and sandhill cranes. Trout dimpled after insects, and farm animals grazed lush, flower-strewn pastures. The two coyotes watched idly enough, but there was not a movement of chicken or chickadee that escaped them.

The ranch dogs took great interest in the newcomers, sniffing Viola through the wire with great slow wags of their tails or, with hackles raised, engaging in urinating contests with Jack. But there seemed to be more bluff and bravado than real hostility. By now the ranch dogs had come to accept any new arrival as just another of their master's long line of strange guests.

I watched carefully to observe any behavioral characteristics that might have been caused by captivity, but outside of a tendency toward nervous pacing on the part of Jack, the coyotes could be any two of a dozen I had observed around the ranch.

Since Viola was obviously pregnant and her time fast approaching, I built her a large, snug kennel. On the floor, in the inner darkness, I scattered a quantity of sweet meadow hay so she could whelp in comfort. I donated several weathered wool jackets I'd worn out during the fifties and sixties, and with these she lined her nest, retreating to the security of her den whenever strange noises bothered her or strangers subjected her to scrutiny.

While my wife remained skeptical about the whole operation, she viewed the contribution of my old jackets with great satisfaction. She brooded a bit over my attentions to Viola, tossing me one of her unanswerable questions such as "Why didn't you worry over me all those times I was pregnant?" Then, as though resigned to the cruel trick Fate had played on her by introducing her to an animal lover, she let the matter of the coyotes drop.

When they arrived at the ranch, Jack and Viola were perfectly content to eat prepared dog food, but I felt that I owed them an occasional short-tailed ground squirrel from the teeming colony just west of the ranch house. These they took from my hand, then gobbled them up and paced the fence for more.

Once they had tasted wild fare they refused to return to their old rations. It was natural food or nothing. No matter how skillfully I doctored a pan of kibble with gravy from our table, they acted without grace by kicking dirt into the pan and urinating on the mess. That done, they eyed me hungrily for more squirrels.

Much as I disliked being a hunter, I became a slave to their appetites, wasting several hours a day and a good deal of ammunition just to curry their favor.

Their eyesight was keen. When I was still half a mile from their pen, they could tell whether or not the hunt had been successful. If they decided that what I carried in my hands was food, they charged back and forth along the fence, often joining in a joyous chorus of howls. At my approach they began leaping, and they caught in mid-air whatever I threw over the fence. If I had tossed in my buckskin glove by mistake, they would have swallowed that.

As an experiment, I gathered a broad spectrum of wild fare, mostly road kills I found on the way back from town. When I laid them out in the pen, they ate the mice, voles, and pocket gophers first, and the pine squirrels last. Whenever they were offered this choice, their preferences remained the same.

Jack, like Don Coyote, was not much of a gentleman. Male chauvinist to the core, he drove his pregnant lady away from food with slashing fangs. Only when he had fed would he allow her to touch a morsel.

But there was a toughness to the lady. Often Viola deserved the treatment she got from Jack. Usually I saw trouble brewing before it happened. Jack would try to ignore her ingratiating advances, and Viola would molest his peace, refusing him his dignity, pawing at him with front feet, backing into him with her rump, trying to force him to regurgitate by licking his muzzle, or tossing a stick in front of the cranky old male in an attempt to get him to play with her.

Jack would put up with just so much, then swiftly and savagely he would attack her, knocking her flat with his broad chest and afterward standing over her, fangs bared, daring that fawning, submissive body to so much as wiggle. No blood flowed, and it seemed to be what she had wanted all the time. And yet he seemed truly to hate her then and to welcome her trips into the shelter so that he could be alone. Captivity was an unnatural state; in the wild he would have gone off on long hunts, joining his mate only when he wanted her company.

But at times, perhaps when he thought no human watched, he could be quite tender. He would even play games with her, usually a tug-of-war with a stick or bone. One favorite game was "king of the mountain." Standing on a rocky outcropping, each would try to shoulder the other down the slope, attempting to keep the summit for himself as the other tried time and again to take it away.

Often it was Jack who initiated a game, guarding a special stick or bone, playing keep away, just as Don Coyote had enticed me with the pop can. Like the Don, Jack was visibly disappointed if Viola refused to become interested. Frequently they simply wrestled and chewed on each other, playing at killing, just as though they were reliving their puppyhood.

Although Viola was too shy and nervous to be friendly, we all took turns courting Jack through the wire. He formed friendships easily, but with only one person at a time. If you were lucky enough to be his pal for the day, he would rub up against the wire to be scratched, smile at you with joy as you approached, wag his tail with enthusiasm, and even lick your fingers. But watch out for tomorrow! That same friend from yesterday might approach, and Jack would growl fiercely, raise his hackles, urinate on a corner post, glare the person down, and do his best to bite him through the wire.

And I, who cared for him and bribed him with baby mice from the wild and big, sweet bore worm grubs from rotten logs, had to take my turn with the rest. One had about as much chance of predicting his daily allegiance as one had of guessing the outcome of a game of "he loves me, he loves me not" played with a worm-eaten daisy. If he liked someone, that person could walk right in and play with him; if he was mad at you, there was no way you could get in and out of the pen without being bitten. He bit two or three people who thought he was always amiable.

Obviously, captivity influenced the coyotes' dispositions, but there was still much to be learned from them. Many a time I had lain in bed in the old stone house listening to what sounded like a pack of coyotes in full chase, envisioning them sweeping down across meadows and over ridges as, by sheer numbers, they swarmed over and engulfed much larger prey. It was an eerie, chilling sound, one that had struck many a rancher before me with sickening apprehension. At the first light of dawn I was always relieved to see cattle grazing peacefully on the meadows, mares nursing their foals, with nary a sign that the terror of the night had ever been.

It took Jack and Viola to set me straight. One morning as I lay abed, watching the first light of dawn make sense of the patterns on the curtains, transforming weird shapes and nighttime dinosaurs into commonplace household furnishings, I heard a wild outcry from the direction of the coyote run. It was the same frightening sound I'd heard so many times before—a huge pack of coyotes rapt in a hysterical orgy of killing.

Hoping to save whatever it was the pack was pursuing, I leaped from bed and ran naked across the meadow. Near the house spring I stopped short in amazement. Every bit of that savage chorus was emanating from Jack and Viola. Instead of a pack in full chase, there were only two coyotes, sitting under their big pine tree, noses tilted to the sky, having a whale of a good time.

It made me wonder how many coyotes in the past had

been blamed for what it sounded as though they were doing, and how many ranchers had taken the time to track down that wild chorus to see what the coyotes were actually up to. In the future I could listen to that music for what it was, just one more movement in the rich and varied coyote symphony.

17

VIOLA HAD BEEN AT Yamsi nearly a month when nervous actions seemed to indicate that whelping time was near. There were physical manifestations as well: She moved a bit heavily, and though no dugs showed through her belly hair, when she arose from a bed of dry pine needles, she left a wet spot where her tail had rested.

I refurbished her whelping house with soft, fragrant cedar sawdust, plus a mountain of clover hay and some torn-up cotton-flannel nightgowns—everything a coyote mother could desire. Then, on one of the days I happened to be in Jack's good graces, I nailed up part of her front door so she could repel Jack's advances, should he take offense at her pups.

Moments after I left, Viola made a tour of the delivery room. I watched from the spidery gloom of the black- smith shop, peering through a flicker hole in the split cedar shakes of one gable, feeling an excitement mount in me, thinking that she had to be the luckiest female coyote in America to be so well provided. She had everything but her own private physician, and if there was trouble, I could probably help.

That evening, with new confidence, she shouldered Jack away so she could take a ground squirrel and fled with it to a corner of the pen, where she laid it down and eyed it quizzically. So closely did it resemble a newborn coyote pup that perhaps she was torn with indecision as to whether to eat it or nurse it. Picking it up by the scruff of the neck, she carried it around the pen at a lope for a few laps, then, when Jack made a move for it, she carried it into her quarters, where she laid it in her nest and curled up beside it.

Two days later, snug in her bed, Viola gave birth to six blind, helpless pups. Insatiably curious, I provided myself with a voyeur's peephole, complete with a black shroud, which increased my night vision and kept Viola from being alerted by flashes of light.

Along my access route I had cleverly screened off one side of the pen with solid panels, but I had reckoned without the coyotes' sharp senses. Where their eyes failed them, their acute senses of hearing and smell announced my presence. However stealthy my approach, I heard rapid coyote footsteps as Jack began to pace the panels separating us. I shrugged and peeked in anyway.

I hadn't fooled Viola either, for my next view of the interior showed her staring right back at me. She propped herself up on her front legs as though ready to flee, but I sat at my knothole without even blinking, and soon she settled back comfortably.

A warning growl from Jack told me that I wasn't fooling them a bit and had better behave.

Once my eyes grew accustomed to the gloom, I could see details of the room. Viola was still panting from her ordeal, warming the pups with her soft body. To the rags of the nest, she had added body hair plucked from her swelling teats so that the pups could nurse more easily. The pups looked like six mature ground squirrels. Even the coloration was similar.

Now and then one of the pups, rooting blindly, would dislodge another from a teat, and the displaced pup, rolling to the edge of the squirming pile of siblings, would right himself and head back to begin a displacement process of his own. Seldom did one pup get away with nursing undisturbed, for the sound of sucking stimulated the rest of the brood into frenzied action.

Now and again, Viola pushed herself into a sitting position to lick her pups, cleaning up puppy indiscretions with her tongue, licking and massaging their bodies as though kneading them to life. It was as though she understood that through such actions of her tongue the excretory processes of a tiny canine are begun and that without such ministrations its chances for survival would be small.

I stayed watching until my body ached. My shirttail hung loose behind, allowing great drafts of cold air up my back, yet I dared not tuck it in for fear the slightest rustle would frighten her. My knees felt every rock, and my feet were cocked at an angle not often achieved by ballerinas. But I toughed it out until darkness stole my vision and the big black dinner bell hanging outside the ranch kitchen summoned me for supper.

Years back, some long forgotten cowboy once described a typical day on our ranch as "one big mismanaged emergency after another." With the coyotes it got even worse. Add to ranch and cattle chores the task of providing natural food for a pair of coyotes, and the effort seemed Herculean.

Predators time their birthing to coincide with the approach of a reliable food supply, so that when the young are making the heaviest nutritional demands on their parents, the hunting is also at its peak. Coming as they had from Pennsylvania, our coyotes were out of sync with what was available in Oregon, whelping earlier than our locals, which generally seemed to coordinate their period of highest need with the emergence of short-tailed ground squirrels or other young prey animals.

While there were a few mature ground squirrels out and about, the arrival of Viola's pups was followed by a series of cold blustery days when the ground squirrels were as reluctant as I to come outdoors. The only ones I saw were a few superalert sentinels appearing for a few moments to scout the weather.

With a moral obligation to the coyotes to provide the equivalent of fifteen ground squirrels a day, I failed to accomplish much around the ranch, but I did end up with a solid appreciation of just how much each predator family on a ranch contributes by rodent consumption. As the pups grew, I would need even more food for them. There were times when I considered releasing them to do their own hunting.

Jack's only concession to fatherhood was to defend his territory with louder growls and more frequent post drenchings. Instead of always competing voraciously with Viola for food, now and then he would act the solicitous husband, standing back to give her first pick of the ground squirrels as I dropped them into the cage.

If Viola failed to get enough, she crawled to Jack on her belly, grinning her white-toothed supplicant smile, then licked his jaws to force him to release the contents of his stomach. Hurriedly, she gobbled up whatever she wanted.

Since the purpose of keeping Jack and Viola was to raise the young and learn from them, I decided to steal two of the puppies just after they had ingested some of their mother's antibody-enriched colostrum milk, and before their eyes opened, in hopes that I would be imprinted upon them as their mother instead of Viola.

After those two pups were under control, I planned to abduct two more and to leave the remaining two pups with the parents as a control. As the six pups grew, I'd have a chance to observe whatever behavioral differences might exist between them. I had no desire to domesticate the pups, only to tame them to the point where I could release them on the ranch and observe their everyday life without their being concerned about my presence.

"Steal some pups?" Taylor exclaimed when I asked his help. "Neato, Pops, but how are you going to bring that one off without losing an arm?"

"Nothing to it. You take the coyotes a bucket of ground squirrels, and while they are distracted, I'll slip up on the blind side of the pen, reach into the den area, and latch onto a couple of pups."

"Won't work," Taylor said. "They'll read us like a book."

"Try it," I fought back. "That's an order!"

Taylor dumped a batch of squirrels into the pen, and the two coyotes rushed to eat. But Viola only sniffed and rushed back to her house. I looked up from my crawling, feeling that someone was watching. There was Viola, with both forepaws atop the panel looking down at me.

"Told ya," Taylor said.

I pretended to be looking at a pair of barn swallows building a nest under the eaves of a shed, then slipped away. Her sad eyes haunted me. If she associated us with the loss of her pups, she'd hate us forever. There was only one way out: to so alter our appearances that the coyotes would not link us with the kidnapping.

"Now what?" Taylor asked, joining me as I meditated.

"We'll sleep on it. Tomorrow we'll disguise ourselves so that they'll never in the world recognize us."

"Humph," Taylor grunted. "You'll never fool those coyotes."

The rest of the family shared Taylor's distaste for the project. The next morning I tried to enlist my wife's aid. "While you're in town today, dear, how about dropping by the vet's and picking up a supply of Esbilac. It's a milk substitute dog owners use to save orphaned pups."

"No way," she retorted. "I suppose the vet's the only one in town that handles it, and sheepman that he is, he'd probably rather flush his supply down the drain than let you have it to save coyote pups."

"He'll never know."

"He'll read my mind and turn me over to the police. By being an accomplice, I'll acquire a criminal record and be drummed out of the bridge group."

"Get the Esbilac!" I snarled, baring my canines and trying what seemed to work for Jack.

"You go. You're good at lying."

"Move it!"

"You're hurting my arm," she said as I helped her into the pick-up.

Being a criminal came hard to her. All she could think of as she drove to town was that we were keeping coyotes without a permit, and by the time she arrived at the vet's, she had forgotten the name of the product.

The veterinarian stood behind the counter and eyed her curiously as she entered. He was so big and burly he could have squashed her like a bug. Behind him on the wall were photographs of his children with 4-H lambs; not one with a calf, pig, or foal.

"I need some of the what-do-you-call-it—that you mix for . . . Dear God!" Her face resembled a garden of wild roses.

"Which whatchamacallit?" he snapped.

"The stuff for orphaned puppies."

"Oh, Esbilac. We got two kinds."

"What's the difference between them?"

His eyes gleamed as though he'd been dying for her to ask. "The regular is for just plain ordinary orphaned pups."

"And the other?"

"The other's laced with strychnine. It's for coyote pups!"

He beat on his knee with one big fist.

Just as she was considering fleeing, a sheepman in a slouch hat and patched Redwing boots came in, picked up a couple of bottles of selenium for white muscle disease in his lambs, and set it on the counter. The vet rang up the sale.

"How's lambin'?" the animal doctor asked.

"Bad. Got some white muscle," the man replied.

"How's yours? Still got that place outside town?"

The vet nodded. "Mine's bad too. Coyotes got a hundred percent of my sheep this year," he said.

My wife had visions of thousands of lambs lying about his corral, their throats neatly slit. "Hundred percent! How many was that?" she asked.

"One," the vet replied. "Just owned the one ewe."

The sheepman shrugged. "Well, if a man's going to be in the sheep business, he's going to lose one once in a while." He left in a hurry, as though he'd just had a premonition of danger to his sheep.

The vet turned back to my wife. "About those pups," he said, then seeing her mental state, he let his face relax into a grin. "Your old man just phoned and told me your secret," he said. "Knew you'd forget the name of the product. He recalled my tellin' him once I'd raised a jillion coyote pups when I was a kid. All I had to play with. Cute as hell, but lock up your old fat hens."

"Did you really lose your sheep to a coyote?" my wife asked.

"Come to think of it, no. But one way to make a customer out of a cowman or sheepman is tell him what he wants to hear."

Back at the ranch I was having my own problems. My wife had departed from the ranch in such haste, she had forgotten that this was the day she was to host some friends for an outing on the ranch. I did my best to entertain the ladies, but my whole repertoire of small talk was exhausted after about three minutes. I was in a hurry to get on with my pup snatching, but it was question and answer period.

"And are your cattle breeding now?" asked a blonde with a face chiseled in alabaster as she stood on one leg like a heron and Kleenexed a bit of barnyard mixings off a suede pump.

I glanced at my watch, and thought of a flip retort about them waiting for dark.

"Tell me . . ." queried a tall brunette, trying to seem interested. "Tell me what you do on a typical day here on the ranch."

I hesitated. "Well, this time of year I usually spend part of the day riding fence."

"Riding fence?" Her brown eyes widened. "For heaven's sake. Isn't it painful when you come to the posts?"

My wife's arrival spared me further anguish.

Retreating to my bedroom, I stripped and dosed myself with some partly evaporated cologne abandoned by some long forgotten guest. Then, since coyotes are often upset by a simple change of clothing on the part of their keepers, I dressed carefully, putting on a pair of trousers left in a closet by my uncle, a battered shirt I'd bought back twice from rummage sales, and a moth-eaten woolen jacket. Over my face I put a section of panty hose, and over that sequined sunglasses. To complete my wardrobe, I commandeered a floppy straw hat half eaten by a Hereford steer and bleached by sun and rain.

"EEEEEeee!" Marsha screamed as I tried to slip down the hall past her room.

The disguise was extreme, but it made me feel better; I seemed to have moved out of my identity. It made the heinous act I was contemplating somewhat easier. I descended the back steps, the domain of our St. Bernard. "Nice Noah!" I pleaded.

"Woof!" Noah barked, taking me by one pant cuff.

"Taylor! Tell your dog to let me go!"

At the sound of my voice, the dog relaxed his grip and sat down, puzzled. I was pleased that my disguise had worked.

"Yipes! 'Scuse me, ma'am," John said as he whizzed down the hill on his bicycle. Making a skidding turn, he fled down a path toward the barn.

I caught Taylor coming out of the chicken house. "Something must have died around here," he said, wrinkling his nose.

"It's just cologne," I said. "No wise remarks. Just get dressed and help me."

"Sure thing." He looked me over carefully. "Which place we going to bump over? The bank or the company payroll office?"

"We're pup-napping. Ridiculous as it may seem, with coyotes clothes do make a difference. I want you to decoy the coyotes away again with food while I reach into the nest. They'll never know me in this outfit."

"Great! That leaves them blaming me! They'll hate me forever, and Jack and I have just started to go steady."

"Then put on your bee hat and veil, and an old raincoat smeared with diesel fuel."

"You're kidding!"

"From now on they'll hate anyone they see wearing these outfits, but they'll still accept us as friends."

Moments later, he came out of the carpenter shop wearing his outfit, slinking along like any other mortified thirteen-year-old boy.

Pulling on a pair of red rubber dishwashing gloves to further disguise my scent, I sent Taylor to the front of the pen with bait, while I slipped around back.

The coyotes raced around the pen nervously as he approached and only quieted when he dumped a pail of ground squirrels over the fence. "Okay, Dad!" my accomplice squeaked, doing his best to disguise his voice.

I reached into the nest, half expecting to have my arm bitten off. The nest was deserted! Nothing there but an old bone gnawed clean. Either the parents had eaten the pups or hidden them elsewhere. Discouraged, I sat with Taylor, while Jack and Viola eyed us suspiciously.

"I've never seen them looking so well fed before," I moaned. "They've eaten the pups, that's what."

"Maybe not!" Taylor said. "Listen!"

From the earth beneath us came the whimpering of pups. Jack and Viola had moved them out of the kennel into a den they had dug between the roots of the lodgepole pine. Somewhere deep beneath us the pups were whimpering for their mother.

The earthen covering produced a ventriloqual effect. "Here," Taylor whispered, pointing beneath his feet.

"No, here!" I said, pointing beneath mine.

Rushing to the blacksmith shop, I returned with a large funnel. Placing the small end in one ear, I crawled over the ground, using the funnel as a stethoscope. It worked perfectly, and soon I had marked the exact source of the sounds. "We've got them now," I murmured through my mask.

"And on our left, ladies, the last two members of a rare tribe of bee-keeping Indians performing one of their mystic fertility rights," my wife's voice said, ending with a little waver and sob, as though I had contrived the whole scene to embarrass her before her friends.

"You and your big idea, Dad. Pete's mom is there. Wait until this gets back to the rest of the guys on the team. I've spent thirteen years in that town building a reputation for being cool. I listen to you, and I blow it all in a minute."

I ignored the audience. "Too late now. Hand me that big flat rock over there."

With an empty tomato juice can, I dug into the soft earth just beyond the fence, working my way between roots until the last bit of dirt caved in and exposed a chamber lined with coyote hair. There, in the dim light I saw a squirm of pups. Reaching down, I picked two babies at random, then carefully placed the flat stone over the hole and covered it with dirt.

"Those are coyotes?" Taylor asked, disappointed. "They look more like squirrels. Are you sure you dug up the right den?"

Rushing our charges to the kitchen, we put the two blind, helpless babies into a cigar box lined with soft flannel rags, then set them on the open door of the oven. Soon they were fast asleep.

My wife had issued one of her firm policy statements that she was too busy raising five children and running a ranch to undertake bottle-feeding coyote pups.

"Who needs her, anyhow?" I told Taylor when we had shed our disguises and returned to the kitchen.

Opening a can of Esbilac, I mixed some with water, stirring vigorously as a pall of fine white powder settled over the kitchen.

"I'm getting out of here before Mom comes," Taylor said, charging out the back door.

I warmed the concoction on the stove until the kitchen began to smell like burning galoshes, then tried my best to pour the hot milk into a tiny doll bottle. My wife and her friends filed silently into the kitchen and stood like pallbearers at a funeral.

"Last Fourth of July in the wild cow milking contest at the rodeo, you got more milk in a bottle than that," my wife said, surveying a growing white puddle on the floor. "Next time try it over the sink."

I awakened the sleeping pups, dashed a little milk on my wrist for temperature, then tried in vain to get them to suck. They chewed each other all over, but refused the rubber nipple. With my big, clumsy hands, time and again I tried to find the mark. Shrugging, Gerdi took the bottle from me, and in a moment a pup was sucking eagerly and little bubbles were rising in the glass.

"When the coyote mother is through feeding the pups," I advised, "she carefully licks them off." The whole group glared at me.

When the babies were finally fed, we compromised on this point by wiping them carefully with damp tissue paper. Beyond food, they needed cleaning, loving, and massaging. Stomachs distended, they snuggled into the softness of their nest on the oven door and slept.

I made the town ladies swear they wouldn't breathe a word about our coyote pups, though I had a hunch such a secret would not be easily kept.

"I hope you appreciate your wife," one of them said to me as they departed. Then, as a parting shot to Gerdi, she said, "If you ever want us to testify in divorce court, say the word."

18

EVERYONE SHOULD HAVE A worry log. There wasn't a week when I didn't feel the need of its calming influence; every week I'd go there for a time to sit and plan my work. I had no need of company; the log was companion enough. When you poured out your troubles to a tree that had lived for three hundred odd years through fire, drought, and winter gales, then spent another hundred on the ground taking all that the forces of decay could throw at it, you felt buoyed up, as though maybe your troubles didn't amount to beans.

Don Coyote had a scent post at one end of the log, and if you sat there quietly of an evening, very often you saw him as he left on his rounds, gimping along maybe and a bit out of balance, with only three legs and no tail, but going about his business in spite of it. Perhaps, being thus encumbered, he tried harder. Very likely, the less he relied on his body, the more he used his brain.

As long as I stayed quiet and blended in with the great gray log, he payed me little attention, but I had only to move and he would go wheeling out of there, using every rock, every fallen log for cover. He would simply vanish for half a minute or more, and then I would see him again, slowed down to a trot now, heading west along the meadow or north up the river.

He would stop frequently to freshen up the boundaries of his territory, although to my knowledge no other coyotes had moved in to try to claim his range. With the population down, there was plenty of range for all.

From my vantage point on the log I could sit and listen to the various families of coyotes as they talked to each other; the nearest group was still about four miles away to the north. In time, they would move closer, of course, for Nature abhors a vacuum, and Don Coyote would have to defend his territory. He was well healed by now, however, and unless the newcomer was a big, vigorous male, I thought he could handle it.

Sometimes I took the little pups with me in their box and fed them in a nest I made for them in the rotting duff atop the log. Blind as they were, I fancied that they could pick up the scents of the forest that would one day be their home.

Once the pups had made the transition from Viola's tiny pink nipples to the black rubber dug on a doll's bottle and our human smells had supplanted those of the parent, they accepted us completely. Let them but feel the vibration of approaching feet, and they would begin to squeak plaintively, nosing blindly for a nipple, upending each other in the process with slow, salamander-like persistence. It wasn't easy for me, this playing at being a mother. Often in their eagerness to nurse, they only fumbled at the nipple, then churned back and forth, missing the mark like calves nursing a cow's knees, sucking every object except the right one. Trying to place a nipple

in the mouth of a squirming, pawing, overeager little
devil, squeaking like a tape recorder on fast forward,
takes a great deal of finesse and concentration, neither of
which is among my strong points. I tried my best, but
the pups had an affinity for the knuckles of my fingers. I
just knew they were trying their best to die.

"Did you burp them?" my wife asked, taking one of
them from me in disgust at the next feeding.

"Did I what?"

"Burp them." She held the tiny pup against her shoulder and patted it tenderly as it tried to nibble her earlobe.

"There!" she said and laid the pup down on her lap.

I didn't hear the burp, and the whole thing might have
been a put-on, but the pup took the nipple eagerly. The
sucking noises it made drove my pup into a frenzy. I held
it against my shoulder and gave it a few awkward little
taps. I fancied I heard a little hiss, and sure enough, the
baby took the nipple.

"Better cover yourself with a towel," my wife advised.
"What runs in has to run out."

Her advice came too late. Already a dark stain had run
down my trouser leg. But I ignored that hot, wet feeling
from my childhood. My pup was actually sucking!

My triumph, however, was short-lived. The pup
pawed the nipple away, whimpered, and groped about as
though looking for its real mother.

"Colic," Gerdi said. "Run him through the burp routine."

This time a hot, wet stream ran down my shirt front.

"Don't be so stiff and nervous," she advised. "Pretend
you're a mother coyote giving milk, and the little pup is
kneading your udder with his tiny paws."

I was ready to try anything rather than have the death
of a pup on my conscience. Funny thing, it seemed to
work. My eyes glazed in contentment; when next I
looked at the bottle, it was half empty. Soon the pups'
bellies were distended with warm milk, and they ceased

squirming, content to doze. With a sigh of relief, I began
to put them back into their box.

"Not so fast," my wife advised. She wiped each one
carefully with a moist rag, massaging their tummies
lightly as though her fingers were a mother's tongue, letting them nestle for long minutes in the warmth of her
cupped hands so that they would come to know her
smell. Once they slept, she placed their box back on the
open door of the oven and tucked them in with a soft
flannel rag.

"Hey!" I pouted. "That's my favorite shirt. My
mother gave it to me shortly before she died."

"It has a hole in it."

"Same hole it's had for years," I sighed. I made a move
to retrieve my treasure, but the sight of the two little
pups snuggling beneath its softness touched my hard
heart. I turned away lest my wife catch my smile of
mercy.

At three the next morning, my wife grumbled as her
bare feet hit the floor. "I've done this for five children,"
she muttered. "That should have been enough."

From the kitchen I could hear tiny cries of hunger, and
I snuggled deeper under my comforter, pretending sleep,
hoping she would take care of it. After all, she was so
much better at it than I. With my clumsy hands I was just
barely adequate. Despite several changes, my clothes now
smelt of sour milk. It was a fifty-fifty proposition: half
down the pup, half down my clothing.

"Wake up!" she commanded, placing a groping pup on
my face and propping a warm, wet bottle against my ear.
Once her pup began sucking noisily, mine latched on to
the tip of my nose and dug his tiny claws into my lip.
Grudgingly I gave in.

"I'm glad all we have to raise is two," she said when
she finally crawled back into bed, having put the pups to
sleep in the kitchen.

I screwed up my courage. "Didn't I tell you?" I asked.

"In a day or two, when Viola's remaining pups get their eyes open, she'll have two and we'll have four. That way, I'll be able to observe any differences in the pups that might arise from leaving them on the mother until after their eyes open."

She groaned. "You get so excited. Why aren't you that enthused when it comes to a three-o'clock feeding?"

I drifted off to sleep, hoping she didn't really expect an answer.

Two days later, once our pups' eyes were beginning to open, I dreamed up another outrageous disguise and pretended to be hunting arrowheads on the hillside above the coyote run. Despite my activities, Jack and Viola seemed to sense that I was up to no good. They left their pile of ground squirrels and stood watching carefully, Jack bristling and hostile, Viola sad and dejected.

By now, however, I had little conscience left. I strode down the hillside, uncovered the rock, and exposed the inner labyrinth. Startled by the sudden light, the pups tumbled over themselves trying to escape. Swiftly, I captured two of them, tucked them in my shirt front, and closed the hole. Once I was out of their parents' sight I peeked at them. Their eyes were open, but they still had the blue cast of babyhood. They peered at me as though a bit uncertain whether I was a friend or a giant eagle carrying them to their doom.

Viola was obviously doing a better job raising her pups than Gerdi and I. Hard as we had worked, these pups were larger and more handsome. Their brown coats were smooth and lustrous, while our pups were ratty, with patchy fur and stomachs bulging like those of fat old men at the beach.

The new arrivals quickly accepted the bottle, but within a matter of days looked as tattered and neglected as the first two.

"What can we do with them?" my wife asked in despair. "We're giving them vitamins, minerals, a special milk designed for motherless canids, and tender loving care, but somehow we're failing in our commitment to them."

I would have worried more had they not been so vigorous. They slept at times, but when they were awake they wrestled and mouthed each other nonstop. I had a hunch that when they graduated to a natural diet they would grow like weeds. Fortunately that turned out to be correct.

Once the pups acquired some mobility, it was possible to set them out on the lawn in the sun and feed two at a time by holding a bottle in each hand.

Many a newborn calf has followed the cowboy instead of the cow; a duckling hatched in an incubator can be imprinted to accept all sorts of things as parents, even robots. So it was with the tiny coyotes. Even the second pair, who had seen their parents, though never in the light of day, adopted us without reservations. These two were slightly more nervous and sensitive to outside disturbances than the first two, but that was the only behavioral difference I could detect. The four pups followed us about as though we were their true parents.

Tummies distended, bowlegged and uncertain, they toddled after us whenever we moved, as though terrified of being left to a predator, ceasing their mewling whines only when they gained the safety of our legs.

Even at this tender age, timidity seemed natural to them. Each squawk of a Steller's jay or magpie from the surrounding pines, each new or unusual sound, made them dash to us for safety. It bothered me that I had no way of teaching them which sounds were friendly or innocuous and which represented danger.

One day Gerdi came out of the ranch house to see me scuttling across the lawn on all fours with the pups tumbling along behind.

"Look out!" I shouted, as I dove head first under the picnic table, with the pups in hot pursuit.

"Quick!" I snarled. "Dive under here with me before you screw up the whole lesson! There's a bald eagle circling overhead, and I'm training the pups to take cover!" She stood pouting and uncertain, so I grabbed her by the ankle and pulled her off her feet.

I crouched silently, mentally trying to pump adrenaline through my system so the pups could smell my fear.

"Are you sure this is the way the parents teach them?" Gerdi asked, not at all fond of the game.

"What now?" Taylor asked, trooping out the kitchen door. "Just think! Only eight more years, and I'll be twenty-one and away from this loony bin."

The pups showed no fear, only licked our faces and pawed at us in play. I crawled out, feeling my years. Overhead, the sky was empty; the eagle had swept on over the ridges to other valleys, so I helped my wife to her feet. Around us the pack chased each other, mouths agape, practicing mayhem.

I stretched out on the lawn contentedly, looking off across sun-drenched meadows. Above me, tree swallows circled and banked, and from their nests under the eaves, barn swallows chittered. With my wife beside me, I watched over the pups as they played at killing each other, just as our children had played at cops and robbers. Basking in warmth, I was Jack and Gerdi was Viola.

"You're licking my arm," Gerdi complained, but I think she understood.

Don Coyote at the rimrock

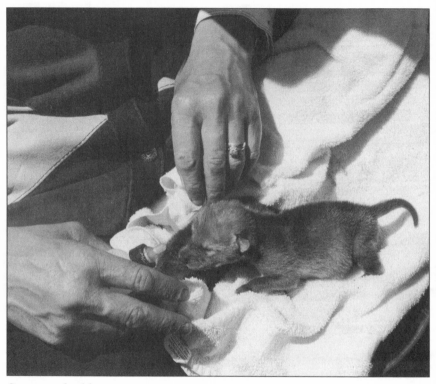

Coy a week old

Benji at three weeks

Coy, Nick, and Benji

Benji pouncing on a grasshopper

Building the dam

Benji full grown

Big Alice pulls the cabin

Coy on the cabin bed

Author, Coy, and a visiting setter

Coy in the snow

Coy having fun

Howling Coy

Author and Coy

Don Coyote waiting for Coy

19

THE COYOTES MADE CHAOS of our work schedule. To survive the demands of two adult coyote appetites plus those of six rapidly growing pups, I had to make some adjustments that would enable me to work the ranch, harvest a load of ground squirrels each and every day, and mother the four pups in my care.

I solved the problem by taking the pups with me wherever I went. They adjusted quickly to riding on the seat of the pick-up, but what they really loved was life ahorseback, nestling happily on a pile of worn Levis, two on each side of the saddle bag. They would peer out at the scenery, ducking down deep whenever anything caused them alarm.

In the beginning, Straight Edge frequently reached around to sniff the pups, cracking vertebrae, first on one side, then the other. The pups would make her straighten up by yapping whenever her nose got too close. Another horse might have bucked us all off and stampeded across the flats, but the mare had been born gentle and had grown even more mellow through the years. Nothing I asked her to do seemed to surprise her.

Since there were no neighbors to whom we could entrust the pups, they came along whenever we had to take automobile trips. On the first rides they were violently carsick, but once they adjusted they would spend the trips in wrestling matches in the back of the station wagon or staring out the windows at the vague world of racing images, most of which had never been part of their experience. Any loud noise would send them diving for shelter.

When they became hungry, they slid like otters over the front backrest and down onto our laps. Sometimes a gas station attendant would stare at them through the windows, and for one brief moment I would tremble lest he report us to the police. At this stage, however, they looked like dogs, and I could easily have passed them off as sheltie pups.

Little by little, my wife came to trust me to solo with them, until at last I took them with me to California on a cattle-buying junket.

"How do they look to you?" I asked on my return.

"Terrible! They look like a pack of Mexican hairless pups with skin disease. But they seem active enough. If they intend to live, maybe we'd better give them names."

"Coyotes don't have names. It's too silly."

"How about Don Coyote? And Jack and Viola?"

I shrugged and accepted the inevitable. After all, the pups had become such a topic of family conversation that good, sensible, identifying names would help us in discussing their behavior.

"How about naming them 'One,' 'Two,' 'Three,' and 'Four'?" I suggested, dreading what the children might come up with.

"Oh, Dad!" Marsha complained. "You're no fun at all."

By an illogical process known only to children, the pups came to be called "Coy," "Boy," "Nick," and "Benji."

If Viola's litter had been born earlier than those of local coyotes and out of sync with peaks in the food supply, I somehow tided them over the lean times until the first baby ground squirrels emerged from their burrows. Suddenly, where there had been two picket pins standing beside a burrow there were now eight or ten. It was a time of plenty, but I was not the only predator working the colonies. Red-tailed hawks screamed angrily at the competition. Goshawks skimmed the meadows, badgers challenged my right-of-way, then backed off to dig themselves into the first available burrow.

One morning, as I lay waiting for squirrels to emerge, the field was visited by a golden eagle that descended a stairwell in the clouds, a pair of goshawks that habitually nest in the lodgepole pine cathedrals near Deer Draw, a great gray owl, two ravens, a red fox, a bobcat, a pair of weasels, and Don Coyote himself, who came along in time to put to flight two fat, shuffling badgers, who hissed angrily but retreated.

The young ground squirrels had not yet become wary and were an easy mark for those who fed upon them. With such a good supply of food, I was able to wean the pups from the bottle and to put out their puppy chow and milk in a dish.

It was Nick who discovered the concoction first. He tasted it daintily, then accepted the food by placing both feet in the milk, as though fording a stream. Now and then he would raise his silky muzzle to check the whereabouts of the others.

It took some time for the rest to catch on, and Nick had the dish all to himself until suddenly the others seemed to sense that they were missing out. Charging up in a pack, they bowled Nick over in the process. He growled his best growl and bristled, but that only convinced the others he was on to something big. First Benji investigated, then Boy. Within moments the newcomers were slurping happily away with little pink tongues, making up for lost time. Nick, stomach distended with gruel, tumbled over on his side in the sunshine and fell fast asleep.

With a change of diet, the puppies grew fast, and their hair covered their bare spots so that they became almost handsome.

The differences between the four pups was slight. Coy and Boy, taken from their mother when their eyes were closed, were calm, while the other two, Nick and Benji, taken after their eyes had opened, were slightly more nervous and independent of me. When all four were playing together, an unfamiliar sound was apt to send Nick and Benji racing for a hiding place in the delphiniums.

The two pups we had left as a control with Jack and Viola had by now ventured out of their burrow into the world, but they were astoundingly different than their littermates. Even though Jack and Viola showed little nervousness at my approach, the two pups panicked whenever they saw me, diving into their burrow at my approach. My main sightings of them were through binoculars, and even then they seemed to sense that they were being watched. I was reminded even more how important proper diet and parental care are for a young coyote, for the two pups were half again as large as those we had struggled to raise.

One marked difference between their life and those of wild coyotes was that Jack and Viola had no real retreat from the pups' demands for food or frolic except by employing discipline, and they exerted that influence early.

The pups soon learned that they could wool Jack mercilessly as he lay sunning himself, growling and chewing at his ruff, tail, muzzle, feet, and ears, but let Jack give a warning growl, and they would back off. If they persisted, he bowled them over in a savage, fangs-bared rush that sent them scurrying backward in retreat. They could nurse Viola, but when she rose to find a spot away from them, they had best not follow, for she, too, could turn into a snapping, snarling stranger.

Jack and Viola gobbled up each dump of ground squirrels without sharing with the pups, though they later regurgitated and sat watching patiently and seemingly bemused as the pups fought fierce, noisy wars over the spoils.

All six of the pups spent most of their waking hours engaged in mock combat, mouths gaping impossibly wide, looking like jaw-sprung alligators as they clamped on heads and necks, wrestling each other to the ground as though killing prey. At those moments the mouths seemed disproportionate to the body, and the teeth white and sharp, but the constant tussling produced neither wounds nor enmities. For all the pups, life seemed to be progressing smoothly.

Tragedy, however, soon broke up the group. One day I heard a fearful commotion in the chicken house some distance from the coyotes' run. Expecting disaster in the form of a raccoon, I rushed to aid the stricken birds. The excited chickens were safe on their perches, but there, hiding in terror in one dark corner, was one of Jack's pups. It had dug past the den in which it had been born and had unexpectedly broken through a dozen feet beyond the fence.

Gently, I caught the poor frightened animal in my arms. It made no attempt to bite, but shaking in terror, hid its head in my shirt and did not struggle.

Jack and Viola paced the fence at my approach, while Jack bared his fangs at me. Releasing the pup in the pen, I beat a hasty retreat so that they could soothe their frightened offspring.

It was over in seconds. Jack took one smell of the cowering pup and killed him instantly. Perhaps my scent on the pup had turned him into a stranger.

I feared for the other pup, but in spite of my doubts, he flourished. Indeed, the patience of his parents was remarkable. No longer possessed of a littermate with whom to wrestle, he turned his attentions to Jack, until that poor old gentleman took to rising to his feet and standing for long periods, elevating his nose out of the pup's reach.

Even though the pup had witnessed the killing, he seemed to have no inkling of Jack's capacity for violence. Again and again I saw him pounce as the huge old coyote lay dreaming and aloof, or even snatch away a bone and run back past his father, taunting him. If he did get into trouble with Jack, he dropped down fawning in submission on his back, one hind leg cocked in the air.

As the pup grew older, it became increasingly nervous and unhappy with the very captivity that Jack and Viola accepted so gracefully. Opening the door to the run, I let it streak for freedom. When I next saw the animal it was half a mile away and still running.

Now and again in the ensuing weeks, we caught sight of the pup out mousing for himself on the meadows, but it was as wild as any wild coyote who had been shot at. It had met Man, hated what it saw, and run away, wanting no more association with him.

If my pride was hurt, I got over it. The other pups filled my life with joy. The lawn surrounding the ranch house and the gardens nestled amongst great lava ledges remained their world, though my wife would have given her eyeteeth to have them play elsewhere than amongst the flowers. Let a blossom cheat the summer frost and bloom, and the pups would chew it to shreds. It was a common sight to see one pup streaking across the lawn carrying a petunia blossom, as though it were a brilliant butterfly, with the other three pups in full pursuit.

It did no good at all to scold them. They cared not one whit whether or not their actions pleased us, and they even seemed to enjoy teasing. Whereas domestic dogs have been bred to obey and try to please their masters, in the coyotes there was no such sense of obedience. Whenever you wanted them to act a certain way or refrain from doing something bad, you had to trick them into conforming to your wishes: If one was into mischief, you diverted its attention; if you wanted a pup to come to you, you hinted that you might have some choice tidbit about your person. Insatiably curious whenever you did something that looked interesting, they would wander over to investigate.

Stronger now and pointy-eared, they growled, wrestled, and played like other canine pups, but their games seemed geared by Nature specifically to develop their prowess as hunters. They practiced for adulthood on each other, creeping, stalking, pulling down prey with paws over the neck and a wide-mouthed bite to the jugular.

In play, they bit softly, as though some ethic prevented them from breaking skin. Often a pup would take my hand in his mouth and hold it gently, as a retriever holds a quail. When a pup was cornered or pushed to his limit, he had only to turn on his adversary with a show of anger, and that snarling display was enough to change the mood back from combat to harmless play.

Perhaps parent coyotes teach their pups certain refinements of the hunt; the rest is instinctive. Coy, Boy, Nick, and Benji took to hunting on their own without any encouragement on my part. Let a housefly land on a blade of grass on the lawn, and they would be after it, creeping forward to pounce stiff-legged. They never failed to show astonishment when the fly eluded their clumsy efforts.

One day a great flying cracker of a grasshopper landed on the lawn and was immediately spotted by Coy, who began her stalk, head low, intent upon her prey. She glanced my way only for an instant, then crept forward again. The other pups ceased their activities and watched curiously. For a moment the little pup paused, nose working as though fighting a sneeze, then, checking the competition to make sure none of us were after her prey, she crept forward.

Nick could stand it no longer and galloped over to investigate. Out of the corner of her eye, Coy saw him coming. It was a long pounce, but she impaled the grasshopper with her front feet and seized the insect in her mouth just as a hurtling mass of pups, led by Nick, bowled her over. In an instant she recovered, and without losing her prize, was off and running, with the rest in hot pursuit. First they wiped out my wife's delphiniums, then a cluster of columbines. Next they obliterated a patch of pansies and bulldozed tunnels through a forest of golden glow. When they finally engaged in a free-for-all amidst the phlox and the alyssum, it was too late for either the insect or the flowers.

As Coy came trotting back to me, leaving the others to sniff the ground for insect parts, she wriggled and smiled with joy. One membraneous wing of the insect, replete with scarlet jewels on a field of yellow, trailed from the corner of her mouth, and she crawled up onto my lap and lay there panting. She seemed to think that she'd caught the only grasshopper in the world and that the others were only wasting time looking for more.

For some moments the others nosed back and forth through the garden, but it was evident they were losing interest. Then Benji made hunters of them forever. Stumbling through what was left of the petunias, he inadvertently stepped on a meadow mouse hiding under fallen leaves. With a yelp of surprise, he investigated with his nose; the mouse squeaked, stood on its hind legs, and bit Benji hard on the tip.

But Benji was not to be deterred. Tossing the mouse

high, he rushed in for a better hold. Nick ran in to help, but Benji dampened his friend's enthusiasm with a growl two sizes too large for his body and a showing of white teeth like the grin of a crocodile.

Pinning the animal with his forepaws, Benji pulled it apart and wolfed it down, while the others sat in a ring and watched for a chance to steal a bit of the prize. Now and again one would advance too close, and Benji would show a forest of baby teeth, daring the offender to take one step closer.

As I watched, the dawning of an idea seemed to come over Boy and pass on to Nick, as though a quiet thought-transfer were taking place between them. Together they trotted off across the pasture to hunt in the taller grasses along the stream. Ears pricked forward, pausing now and then to drink in odors, they were suddenly hunters like all those coyotes who had gone before them. From that moment on there was an independence about the pups. They needed us less, and when not sleeping or playing, they spent much of their time hunting for prey.

The sight of her cherished garden put my wife in a cantankerous mood. Together we tried to resurrect the broken plants, tying some to stakes, clipping others back. Although, with some patience, she had been able to teach the dogs to honor her domain, the coyotes had nothing that would pass for a conscience, and stayed away from her garden only if she stood there with a broom; when she left, they descended upon the area immediately, trying to figure out just what she had been trying to save.

Columbines were a special temptation. Let one gaudy blossom catch their eye, and they would paw down the whole plant and stand on it, stripping off each flower as though it were a butterfly. Only when Nick was stung on the nose by a yellow jacket did he leave columbines alone and develop an aversion to flowers.

20

THE SWITCH FROM MILK to solid foods, supplemented by wild morsels from field and garden, had a marked effect upon their stools, which changed in form and consistency from pancake batter to little pig sausages. New hair with a lustrous bloom covered the sparse velvet pile of puppyhood. They looked like coyotes now and had caught up with Viola's pup in size and condition.

Each of us had had a hand in their care and feeding and felt a special sort of parental pride that they had flourished. We also took pride in the way the pups accepted us as friends and trusted us.

At six weeks of age it was time for their first distemper-hepatitis shots, and I dreaded the ordeal. I picked up the vaccine in town, then did nothing, until fear that they would sicken and die spurred me on to action.

I took each pup off into the woods separately and lavished a great deal of petting and scratching on each animal, hoping to be forgiven, but to no avail. The shots took each pup by surprise, and for a whole day they hid under the wild rosebushes north of the house, nursing sore necks and grudges against me. At last, some toasted cheese sandwiches lured them back to our world.

First Coy and Boy forgave me, then Nick and Benji. Soon they were all playing about the yard, but there was a watchfulness about them that was new. Whenever I approached, they inspected first one hand then the other to make sure I wasn't carrying a syringe. I was reminded that these were not domestic dogs, but wild animals.

Of the pups, Coy was the most loving and gentle, and by spending more time under foot, got more than her share of affection. But being less timid, she also got into more trouble.

Let a night wind set the clothesline ghosts to dancing, and all the pups would sit at a respectful distance and watch except for Coy, who would plunge in to scatter shirts, sheets, and long-johns about the ranch. If we gathered a herd of bawling cows and calves and drove them toward the corrals to work them, three coyote pups would hide in the barn, watching through cracks, while Coy would sit proudly in the gateway, taking our noisy curses and fist-shakings as tokens of endearment until, eventually, someone galloped up through the herd and sent her flying under a corral fence so the cattle could move on past.

If we had a picnic on the ranch lawn, dogs and coyotes would wait patiently just outside the fence, too well mannered or cautious to do more than wait for tidbits to be flung to them. But Coy often crawled right onto people's plates, and many a choice dinner steak disappeared while the diner was busy in conversation. Her endearing, lip-smacking smile and supplicating approach meant only that she was putting a guest off guard and working herself into position so she could snap up a choice piece of meat and rush off with it.

Coy never met a dog she didn't like, and her effusive greeting turned even the most menacing grump of a strange canine into a new pal. Let the worst old coyote-chaser approach, astonishment written all over its face that here was a coyote facing up and not running, and Coy would exude charm. The hackles standing straight up on the dog's back would flatten, the stiff, hostile tail would suddenly start a great slow metronome beat, the play syndrome would start, and Coy would be off, running three circles to the dog's one.

One of the regulars on the ranch was a huge porcupine named Toby, whom Ginny had rescued from coyotes dining on its mother when it was still a baby not much larger than a coffee mug. With puppies, geese, kittens, fawns, and baby sandhill cranes on the lawn as play-mates, Toby felt at home with all manner of animals, but we introduced the rambunctious coyote pups slowly until the porcupine became well acquainted with them.

Toby was independent. When he fancied tidbits from the kitchen, he descended clumsily from some roost tree near the house, rattled the kitchen door, knocked the lids from garbage cans, and kept up a constant mutter until we let him into the kitchen, where he would sit up on his haunches and eat fresh peaches, cherries, or strawberries. He would hold the fruit tightly in his front paws and nibble away with his long, curved yellow and white teeth until the spilled juice made a rivulet down his naked black chest.

Many a time we returned from town after dark to find him working a slow circle around the outside of the house, banging on doors, rattling windows, muttering porcupine obscenities as he stood on his haunches peering about in his weak-eyed way to see if anyone was bringing him a fresh peach. For the safety of the house, we usually let him in and found him a treat. Once the coyotes had been accepted as regulars, dusk often found Toby, the porcupine; Noah, the St. Bernard; the coyotes; and a handful of cats all side by side at the same dish.

In vain the pups tried to make a playmate of the porcupine, but he was one of Nature's clumsiest creatures, and while he would occasionally give a playful little bounce, there was not much more action in him. The coyotes would dash circles around him trying to tease him into giving chase, but Toby only sat on his haunches to get a better view, then, if things became too boisterous, he shuffled off for the peace and quiet of a tree branch.

No one thought to teach the pups that there were other porcupines in the world beside our Toby, and that they would not be friendly. One day Coy came dashing in from the Big World to find a porcupine eating clover on the front lawn. Mistaking it for Toby, she went dashing in to tease. Just as she skidded to a stop, the porcupine puffed up its quills in fear, turned tail, and belted Coy right across the nose with a tailful of quills.

With a yelp of agony, the startled coyote pup fell over backward, pawing at her nose. Her face was a blizzard of quills, and the more she fought, yapping and chewing, the deeper the quills became embedded. I called to her, and she rushed over immediately and crawled into my lap.

It takes a strong man to hold even a small dog when pulling quills. Coy's sensibility went beyond that of a domestic animal, and she seemed to know that I could help. At first she whimpered with pain as I took pliers to the quills, but she did not try to struggle or bite and managed to hold her nose still until the last barbed arrow was out. When I turned her loose, she ran off snuffling and sneezing. As Toby came shuffling around the corner looking for his wild friend, Coy gave him a wary look, as though she blamed him for her ordeal.

Daily the learning process went on. It was difficult to

determine who learned more, the pups or the human family around them. One day all four pups vanished, and we wandered about for hours, listening in vain as our shouts mocked us from the forest. The yelps of joy that usually came back to us were no more, and the landscape was as silent as though the pups had never been.

I saddled Straight Edge and rode circles about the south end of the ranch trying to pick up tracks, fearing that they had strayed and were now in some strange place, huddled together for comfort, homesick, hungry, and miserable. Now and then I would stop the old mare and call, but there was no answer.

Once, in the distance, I saw Don Coyote hunting mice and wondered if he had found the pups hunting his territory and had killed them. The Don, frightened by my calls, turned tail and ran off into the forest, leaving me feeling more abandoned and alone than ever.

I returned home to find my family grouped around the mouth of a twelve-inch irrigation pipe near the ranch house. Noah ran down the hill to me and barked, then ran proudly back to the group. It was he who had discovered the lost pups and brought help.

"They're down in the pipe," my wife said, "and they won't come out."

"Worse," Taylor corrected, wrinkling his nose. "There's a skunk between them and the opening, and he's plenty mad. Hear that? You can hear him stamping his hind feet."

I leaned down toward the opening. Far down the pipe I could hear a pup whimpering mournfully. The odor of skunk was overpowering.

"How about turning on the pump?" Ginny asked. "It would just take seconds, and they'd be out, skunk and all."

I thought a moment. "It might work. But coyotes are such nervous animals the trauma might be too much. Our best hope is to figure out just where the skunk is and try to remove that section of pipe. Once the skunk is out, the pups will come out on their own."

Moving in a body on our hands and knees, we finally agreed on which section to move. The skunk's stampings in response to ours sounded like the muffled beat of Indian drums.

"Now," I told John and Taylor, "just slip this length away from the others, carry it off into the pines, and release the skunk. It's that simple."

"You do it then," said John.

"I already have a hernia," I replied. "Besides, what could happen? It's just a skunk."

With the help of the boys, I slid out the pipe and smiled with satisfaction as we heard scratching within. We had hardly gone thirty feet when a large black and white skunk dropped out of the end of the pipe and went shuffling off into the pines.

"Good riddance," I said. "Now boys, let's just upend the pipe and lean it against the crotch of that tree."

As we swung the pipe up into the air, there was a wild scratching, and four more skunks dropped at our feet.

"Freeze!" I whispered to the boys. "Maybe they'll think we're a grove of trees."

"Let's pretend we're deer," Taylor said, "and run like hell!"

"Stand fast!" I commanded. "I know skunks."

We might have escaped a drenching had it not been for the pups. Freed from their prison, they rushed out into daylight and, seeing us nearby, came galloping over to greet us. All four skunks raised their tails and cut loose, drenching us.

The pups caught a blinding charge right in their faces. Rushing down the hill, with the pups staggering along behind, we plunged into the icy waters of the house spring. As if playing follow the leader, the pups dove in after. They lay with little more than their noses protruding from the muddied water, while we tried our best to do the same.

21

THE RICH BLACK PEAT muck steamed in the sun as I eased the wheel tractor and grass seeder back up on high ground and sat for a moment envisioning a lush expanse of green once the new seeding of Reed's canary grass germinated and sprouted. To me it was more than just a field of highly productive grass: It represented my first crop planted on my own and symbolized the takeover of my generation from the preceding one, with new ideas and enthusiasms supplanting the old conservatism.

"Be enough tonnage of hay from that new field, we'll never have to worry 'bout wintering cattle again," I bragged to my wife.

Her eyes showed the vaguely worried look they always get when I become enthused—some sort of déjà vu, as though somehow in her marriage she'd been down that road before, going toward a reality never quite up to the dream. But she said nothing. The money and time had already been spent, the old native swamp grasses turned under forever, the new seed planted in the warm, wet earth. Nothing to do now but wait and see.

I turned the key off, and the wheel tractor sputtered and died. Across the way, on the far end of the field, I could see Don Coyote inspecting my work, perhaps wondering why anyone in his right mind would destroy one of the best fields on the ranch for hunting mice and turn it into a patch of bare ground. I smiled, thinking that one day the coyote would be surprised, that the canary grass too would attract game and furnish good hunting. But Don Coyote would also have to wait, leaning more heavily on other areas of his territory.

The big iron carburetor I'd taken off the old pump engine over in the Calimus field had lain underfoot in the blacksmith shop for nearly a month while I did my farming. Soon the last trickle of melted snow water would seep into the bottom of the canyon, and without a pump the yearlings would have nothing to drink. It was high time I got out my welding outfit and repaired it.

In the carburetor's inner workings, I found three spiders, five beetles, and a mummified mouse. No wonder the engine had struggled to run. I cleaned the parts with a bath of solvent, welded a crack in the casing, and reassembled the unit, proud not to have a single nut or part left over. Gathering my wrenches, I loaded the coyote pups into the pick-up and drove over a few ridges to where two parallel cow trails left the main road and led down through ponderosa pines to the pump house.

A warm summer rain had just ceased, but a white wraith of a cloud hung over the meadows, pinned to the sky by sharp spears of lodgepole pines along the meadow edges. Glad to be out of the pick-up, the coyote pups scattered like flushed quail, spreading out through a colony of ground squirrels, rushing headlong forward, catching nothing where a little artful hunting might have filled their bellies.

Benji cornered a golden mantled ground squirrel atop the engine itself, and I had to pick up the coyote bodily and carry him out onto the meadow so I could work. He gave me a look of disdain for spoiling his hunt, but when I returned to the pump house, the ground squirrel made a dash for freedom out the door and into the paws of the coyote, at which point I was accepted back into the pack again.

As the other pups chased Benji for his prize, I busied myself with my wrenches. With its two giant flywheels and slender single-cylinder body, the engine resembled a small locomotive. What had taken me a month to get around to, now took me ten minutes to fix. On the first crank, the monster fired, sending blue donuts of exhaust smoke winging out over the neighboring corrals. The single rod and piston drove back and forth, the flywheels gathered momentum until they became a blur, the wide leather belt slapped rhythmically against the pulleys, the pump shaft moved up and down in the well, and an intermittent trickle of cold water splattered on the floor of the galvanized tin storage tank. A thirsty cow could have kept ahead of the pump.

With time to kill before the tank overflowed, I called the pups in off their hunt and strolled over to the old buckaroo line cabin at the edge of the little meadow. Vandals had stolen the wood stove, wrecked the furniture, and broken out the windowpanes. One item survived—a cowboy chair that some ingenious old-timer had whittled and fitted from a couple of pine boards.

It was a surprisingly sturdy affair. Leaning against the clapboard wall of the porch, I sat back lazily in a puddle of sunshine and watched as the barn swallows wheeled and chased over the meadow and the pups hunted amongst the tall grass beside the shack.

Now and then a pup would chase a swallow's erratic shadow across the grass, but they were quick to learn

from futile efforts and found the whistling, impudent
ground squirrels more practical prey.

Unaccustomed to spare time except that which I re-
served for my worry log, I closed my eyes and tried to let
my thoughts drift lazily. I focused upon my childhood in
northern Michigan, trying to remember the sounds the
breeze off Lake Superior made in poplar, birch, and ma-
ple, but I had been gone from there too long.

A recent memory teased me. The last time I'd been
over on this meadow I'd been excited by an idea. Whatever it had been was gone. With me, ideas were cheap.
They buzzed around my head like face flies in autumn
haunting the sunny side of a barn, then just as quickly
were gone with the frost.

Bang! went the pump engine. Bang! Bang! Bang! The
exhaust left a blue haze like a lake over the bottom.
Water! A lake! That was it. Before Jack and Viola had
distracted me, I was excited about building a dam at the
mouth of the canyon and creating a lake in this bottom,
to store the melt water that rushed down from winter
snowfields in the hills—a lake that would make the well
and the old Fairbanks Morse pump and engine obsolete.

Had my sensible wife been there, she would have seen
the excitement and moved to stop it before it burst into
flame and consumed the parent and everything around
it—like a forest fire that starts as a tiny spark, becomes a
flame, then turns into a conflagration, creating its own
winds and racing out of control, leaving nothing in its
wake.

But she wasn't there, and with luck I could again find
that Indian fishnet weight and also the high-water mark
of the prehistoric lake. With the four pups romping at my
heels, I left the engine to run itself out of fuel and moved
down the drainage. In the thickets the raindrops still
clung to the branches, drenching me as I passed, and everywhere was an air of gloom. The pups seemed to sense

an ancient presence and walked on my heels for safety, growling on occasion at nothing in particular. Stark gray branches of dead trees reached out like hands to clutch my shirt.

Everywhere huge crakes of lodgepole pines, long dead, were lying one on another as though they had perished in some great blight. I walked on trunks, leaping from one to another, passing over the ruin of branches instead of fighting through them, leaving the pups to find their own way through the tangles.

At the end of the valley the thickets ended in a gradual transition from lodgepole to ponderosa pine, forming an open wood. Searching, I found the fishnet weight where I had laid it on a log, and then the watermarks. I hadn't been dreaming! At the bottom of the canyon I happened upon an old horse liniment bottle of clear, pinkish glass, its rubber stopper still intact. Filling the bottle half full from a remnant puddle of snow water, I replaced the stopper. Mosquito larva and small organisms peered back at me as I held it to my eye, leveled it, and sighted along the water lines. It was no longer just a bottle, but a surveyor's level, primitive perhaps, but reasonably effective.

Using the bottle, I engineered a shoreline back through the thickets. It was slow, hot going, since the visibility was often less than thirty feet, and I had only my pocket knife with which to blaze sight markers on the trees. I leveled the bottle, sighted over it to a spot on a tree, then went to mark the spot with my knife. Placing the bottle on the blaze, I shot ahead to the next sighting.

All I'd really intended to build was a small livestock water impoundment, but my crude surveying instrument kept leading me farther and farther up the valley as I followed the potential high-water level of a much larger lake. When, at last, I worked my way out of the brush at the far end of the valley, I stared about me in astonishment. Unless my primitive surveying methods were way off, the lake had a potential of being nearly a mile long and half a mile wide. I walked back to the buckaroo cabin and sat down in the old cowboy chair to contemplate the project.

A dam! What did I know about building one? The biggest one I'd ever built was a sod dam in an irrigation ditch. How much would it cost, and even if it was cheap, how would the ranch budget come up with the money? Did I farm out the job or somehow, by blundering along, could I construct it myself? And what of the ranch? Such demands on my time could only add to the already staggering burdens I placed on my wife and children. Yet if I waited until I could afford it, I'd be too old to enjoy it.

Reason on one side and innate stubbornness on the other. I liked being a dreamer; on the other hand I had no patience with those who only dreamed and could bring nothing to reality.

There was a sadness building in me along with the excitement, the same aloneness I'd felt as a boy when I'd experienced something special and had no one with whom to share it. My family? Drylanders all of them, imprinted with the early morning whicker of horses. I, however, from my first days had known the sound of water lapping against the shore of a northern lake. The cry of a loon does something to a child's soul, and he can never be the same again.

While I was ruminating about the lake, Coy ceased playing with the others and came trotting up the slope, pink tongue hanging loose like a petal on a spent tulip. She plunked herself down beside me and began to chew at a flea, biting her fur into a white froth with tiny, rapid bites of her front teeth, then finished off by scratching the area with one hind foot.

High over Calimus Butte to the east, a great golden eagle descended the heights in a long glide, traversed the valley in moments, dwarfing my project, and was gone over the ridges to the west. The pups pushed me backward off my chair, biting my hair, pawing me for atten-

tion. Jaws wide, they appeared to me, their prey, to be all
mouth and teeth. Coy licked my face, but then stole my
straw hat and raced down the hill with the others in hot
pursuit. There was a tug-of-war at the bottom of the hill,
and each pup ended up with a piece. Across the flat, I saw
Don Coyote come out of the woods to watch, sitting on
his haunches as though wondering what all the commotion was about. He rose to his feet and started toward
them, then spied me and vanished quickly into a tangle of
lodgepoles.

Looking up at the round dome of the butte, I wondered if the spirits of the mountain knew or cared about
the dramatic changes I contemplated, or remembered
back through time when an ancient lake had shown blue
on the valley floor. If I fought against Nature or made
changes displeasing to those spirits, whatever I could
build would be temporary. The mountain had the power
to wash away the grandest edifice I might construct, just
as it had destroyed a lake ages back in time. As I looked
toward the mountain, a feeling of warmth came over me,
and I felt that I would be allowed to rebuild the lake that
time had erased.

The next days were passing comets, and the nights
centuries. The only dam contractor I found came,
looked, was properly impressed, and assumed that the
size of my dream meant that I was also wealthy. He bid
over a hundred thousand dollars for the job, an amount
far beyond my means. If the dam was to be built, I'd
have to build it myself. I, who couldn't even read a grade
stake, would have to learn about earth-movers, vibratory
sheepsfoot compactors, push cats, chimney filters, core
trenches, spillways, grades, slopes, freeboards, cambers,
and the like.

I hired an engineer to do the planning and make blueprints, then filed for a construction permit with the state,
claiming a degree in "Eng." from the University of California at Berkeley, hoping the state engineer would think
it meant Engineering intead of English. Before I knew it,

I was hopelessly involved in building a dam 60 feet high, 250 feet thick at the base, and a quarter of a mile long.

"What's going on over at the Calimus field?" my wife asked one day after I had spent six consecutive days there with the pups.

"Not much."

"How much?"

"Little old project is all."

"Like what little old project?"

"Like building a little old stock pond at the lower end of the field to save pumping water."

That made good sense, and she was about to drop the matter. Then suddenly she looked at me curiously. "Like how big a little old stock pond?"

"Not very big. Like maybe four hundred acres. Just restoring what was there once a few years back."

"How many years back?"

"Well, maybe ten thousand, give or take a few hundred thousand."

"What will it cost?"

"Contractor estimated about a hundred thou. The kids and I can build it ourselves for next to nothing. Buy some old wrecks of earth-movers. Sell them when we've finished the dam. Darn good experience for the children. They'll talk about it all the rest of their lives."

A great weariness came over her. For years, every time I'd become preoccupied with one of my schemes, she'd had to do my work for me—as with the coyote project, where I'd had the idea but she manned the milk bottles and cleaned the yard.

"I'm not going to drive a tractor," she said. "And furthermore, Marsha and Ginny are not going to drive tractors. They have callouses enough for girls."

"I wouldn't dream of getting them involved," I said.

As I gathered the pups around me for a walk to my worry log, Coy reached up and kissed my jaw, coyote style, as though she sensed my solitariness. There was something about Coy: She always seemed to understand.

22

I WAS ENGROSSED IN the coyote pups and the lake, but I hadn't forgotten the planting of canary grass. I checked it daily, watching the land turn from a muddy black to the lush emerald green of a hayfield. Within weeks it was up to a horse's knees and still growing fast. Even the county agent was impressed. Having suggested the grass as a way to increase my hay tonnage, he took a special interest when the experiment appeared successful.

With my blessing, he set up a special tour of my planting to show other ranchers how smart we were, but the morning .of the day they were due to arrive, a summer frost turned the whole verdant planting black. All the long hours on the tractor, all that money spent for seed, fuel, and fertilizer wasted. And all that enthusiasm misspent.

I sought out my worry log, trying to tell myself it was all the fault of the weather and had nothing to do with me, but deep down I knew the error had to be mine. It was time to rethink my whole relationship with this valley and work with Nature instead of running headlong into her. It was all too easy to put the blame on forces beyond my control when, in reality, I was the culprit.

Head in hands, I thought back on tales my uncle had told of the valley as he had first seen it just after the turn of the century: streams running down the draws; grouse by the hundreds in the willows; waterfowl blackening the sky; in the river, trout so thick at times it seemed one could walk across their backs; grass fanning one's stirrups; cows so fat you could have played poker on their backs.

If I dismissed these descriptions as nothing more than a bit of story-telling on the part of an elderly man, what of his stories of later years, many of which I could confirm because I had been there?

I'd seen grasshoppers piled to the eaves of buildings, traveling so fast and thick across meadows a cowboy riding through them risked getting bucked off or losing an eye. I'd seen those same 'hoppers filling the ruts of roads so that tires spun as though on ice. And I'd seen the killing frosts in summer, the hordes of ground squirrels, the dry years when grass withered with frost and drought and seemed to blow away, and the year when the meadow mice turned the country to ruin. If the old man said that on the seventh day of July, 1932, it got down to twenty-three above zero, you could pretty much believe it. He had that kind of mind.

Maybe changes in the land were inevitable when you went from grazing Nature's way, by fire and wildlife, to forage utilization, by haying and grazing cattle. There was nothing shabby about the way the old man had done things. He was a good grass farmer, growing clover and bluegrass where once there had been only native plants. He'd drained the marshes, aerated the soil, and left a proper residual stubble of grass for protection of tender shoots in the spring. If the best fertilizer on a ranch is the footsteps of the owner, his ranch should have prospered. But sitting here on my log, I began to wonder if the old man had done too much too well.

I could close my eyes and hear the old man railing at me to do things his way: "Get that water on and get it off. Go on with every drop you have; never divide the head. Fill the dams, then flush them out!"

Every mile or so down the valley, a low irrigation dam picked up drain water from higher pastures, and scattered it back out over the valley. He kept the water moving, and the moment it started to turn the color of buckaroo coffee, he turned it away and started over.

"Hell, thet water's just gettin' some noodles in it," Homer used to protest. "Don't let it get away!"

Had Homer been on to something? They were all dead and gone now, those ranch-wise old men I'd ridden with—Homer Smith, Slim Fields, Ash Morrow, Ern Morgan, Ernest Paddock, Al Shadley—and no help to me now except that, having spent years listening to their tales, I was a reservoir of their experiences. But how sound were their observations?

Sitting on my worry log, I became aware of some differences between me and the ranchers around me that had been growing for some time. Since Don Coyote had come into my life, my attitudes toward the land had changed. What made me so different now from my neighbors was that they figured we humans had dominion over the land while I felt we had a responsibility for it—for the soil; for every plant, bird, and animal that shared this planet with us; for the rivers, and for the air.

I hadn't pushed the water as well as my uncle, but tended lately to let it warm under the sun before I used it elsewhere down the valley. By my uncle's standards, the irrigation dams along the valley needed a good flushing out; it was time to replace the water with clear, frigid spring water from the Williamson.

I rode Straight Edge to the first of the impoundments and peered down into the murk at a wild confusion of pond life about to be wiped out by my actions. The water was a thick soup of billions of tiny creatures all headed pell-mell for nowhere, a suspension of living entities, like dancers on a crowded floor rushing to each empty spot only to find it jammed full when they got there. And on the floor of the ponds was the detritus of the dead and the dying and their cast-off garments, all adding richness to the marl, releasing borrowed nutrients to oncoming generations. It was the brew of life!

Straight Edge paused, expecting me to open the headgate, but a new stubbornness stayed my hand. Instead, I rode off, letting the water take a few more days of sun, encouraging the brew to thicken.

It was like having a day off. I rode on down the valley, no longer moving methodically from dam to dam, but drifting, enjoying. The red-winged and yellow-headed blackbirds sang from song posts among the reeds; Wilson snipe winnowed from the heights; horned larks tinkled and tumbled down a faultless sky; rails cackled; sandhill cranes ratchetted; hawks and eagles screamed; willets, phalaropes, and terns dove at me and scolded; nighthawks boomed; swallows graced the air; coyotes moused in the grasses. I saw them all as part of a totality. Whether they moved in water or in air, what did it really matter?

The county agent had been wrong about Reed's canary grass for my valley; maybe others before him had been wrong in their assessments of the land. But it takes guts to try to change a system. To embark on a gamble when your financial situation borders on the desperate requires a lot of courage, maybe more courage than I had.

In the next weeks I spent more time on my worry log than I did on Straight Edge. With the pups chasing chipmunks amongst the ruined fibers of the ancient tree, I tried to concentrate on the land and its problems.

Our irrigation water rushed from underground caverns

at fifty degrees or colder, too cool to sustain much plant life. The first ponds were solar energy basins, warming the waters a little before they coursed off across the land. I had never looked at them as such, but now I realized that anything I could do to exploit them, by enlarging them or holding back the water to warm it, would increase production of grasses down the valley.

The mountain sun was fierce and strong; it was the nights that limited production, for the thin air was quick to cool. We were at 4,600 feet, and I couldn't change that. I envied lowlanders in the Klamath area who could raise apples and berries. I couldn't even raise the trees and bushes, let alone the fruit.

But, then, if you didn't have frost-fragile plants, your crops would freeze at night. Even in the Klamath Basin that year it had been a cold one. If I was to survive in that valley, I realized that I had to learn how to change my climate, to manage the land so I wasn't quite so vulnerable to cold. I didn't realize that I'd already taken the first step.

When I was in Klamath a few weeks later, I passed a farmer friend whose face was long as a mule's. Even at that lower elevation, frost had wiped out his barley, and the bank was making him sweat. I could hardly wait to get home, for if he'd suffered a killing frost down there in the banana belt, I feared my ranch must be a devastation of frosted grass. And the weatherman had predicted that that night would be even colder.

As I arrived back at the ranch the next morning, a north wind from off Yamsay Mountain was just chasing the last wisps of fog away. The meadows lay green and sparkling with dew, as though frost had never touched the valley at all. Standing in ponds, sloughs, and reservoirs, my water acted like a giant radiator and all night long had been releasing heat up into the air.

"My garden froze last night," my wife said, riding up beside me as I wandered with Straight Edge down the valley. "Wiped me out; I'm never going to plant anything again."

I smiled at her. "Look out there and learn," I said, nodding toward meadows that could have made the most brilliant emerald seem dull by comparison. My new way of letting Nature take over was paying off. Out there something nice was happening.

23

I NEEDED BIG ALICE to use in building the lake, and as though he realized I was about to steal his castle, Don Coyote went slinking off through the pines at my approach. Had he possessed a tail, it would have hung between his good hind leg and his stub.

There was no way I could tell him that I'd lain awake most of the night wrestling with my guilt. "Sorry, Old Don," I called after him.

Back to work, Big Alice started with a thunderous roar and a belch of black smoke which capped the hill like a rain cloud. The grass under the tracks had yellowed and vanished, leaving two parallel strips of rust resembling pure iron ore. As the tractor clanked off, I glanced back at the den area, well swept by constant use. In an adjoining burrow, spiderwebs trembled in the sudden wind, then collapsed entirely as a toad, blinking in the bright sun, leaped for the hole and pulled the curtains in with him. Moments later I nosed the tractor over the first ridge and headed it for Calimus.

Big Alice purred smoothly and even roared like a lion as the mountain grades began to challenge her engine. I prayed that she would last the job. There was much I needed her for. First, the brush had to be removed from the lake site. After the loggers had taken out any merchantable timber, there would be days of piling and burning slash. Then the dam site had to be pared down to bedrock to satisfy the state engineer. After that, her dozer blade would have to lay bare, from their overburden of four feet of volcanic pumice, the beds of clay sand to be used in the dam. Last but not least, the big tractor would have to serve as a pusher for the huge rubber-tired scrapers I had yet to locate.

For a long way up the ridge, Don Coyote followed Big Alice, perhaps to check on where I was taking her. Halfway down the slope, he seemed to decide digging out a flying squirrel hoard was more important than following, and when I saw him last, he was making the rotten wood fly. Evidently he caught up with the tractor at the lake site, for the next morning there were coyote tracks in the dust around the big machine, and each corner had been washed clean of dust where he had restaked his claim.

The plans for the dam had been drawn up by a local engineer and approved by the state, so the next step was to acquire some earth-movers—decrepit enough to be reasonably priced, but in good enough condition to last the job.

My first acquisition was a Le Tourneau C, an electrically operated earth-mover of uncertain vintage, for what I then considered an incredibly low price. My spirits wavered momentarily when the man seized the money and dashed for the phone, where his first words were "Hey, Gladys, you'll never guess what I just did! Pack your bags, honey. Pack your bags!" Apparently the big old monster had been looking for a home for some time.

It was our agreement that the seller teach me to operate the thing, but that education consisted of starting the engine for me atop a mountain, shoving me up into the driver's seat, disappearing for a moment to find a bathroom, and never returning.

There was no steering wheel; instead, the machine operated by means of several electric motors and switches, most of which took spells of not running at all. If you wanted to steer to the right, you flipped a certain switch; to go left, you flipped another.

I sat atop the hill with the motor running, trying out a few switches, gingerly raising and lowering the carriage a few times, then set off, thinking how easy it was going to be. One moment later, as I went over the lip of the hill, I felt like a student pilot on his first lesson who, as his plane becomes airborne, looks down to see the instructor running along on the ground below.

I have since been told that the C's or Cougars, as they were called, probably killed more operators than any comparable machine in history and came to be banned on most jobs. The great banks of platinum electrical points on my controls were so corroded that sometimes they made no contact at all, and at other times strange things happened.

I was headed down the mountain when the engine itself quit, leaving me with no brakes, no steering, no way to lower the scraper blade. The big wheels racing on each side of the driver's seat were just waiting for me to try to jump. By the time I hit the bottom of the hill, the machine and I were just a blur. Crossing the main highway, I took out three mailboxes, a five-wire barbed wire fence, and an irrigation wheel line before finally coming to stop in an alfalfa field.

Fortunately, the drama was witnessed by a friendly mechanic on his way to town. He stopped his pick-up, drained a gallon of rust flakes from my diesel tank, cleaned the filters, and sent me on my way again.

When you don't need machinery, you see it along every rural lane, all with "For Sale, Cheap" signs upon it. Now every machine I tracked down had been sold the week before at a price the former owner invariably described as a "steal" or a "give-away."

I ended up attending auctions as far away as Los Angeles and came home with another earth-mover little better than the first, plus a compaction machine known as a vibratory sheepsfoot, powered by a cranky diesel engine that, when it felt like running, made the whole wreck shimmy and shake like a go-go dancer with the heaves.

Lined up at the dam site, they were an imposing array of rusty metal. "Know what this looks like, Dad?" Taylor said, eyeing the machines. "Like a museum display of the early history of the construction industry."

"Who's going to drive them?" my wife asked, trying to clutch all five of her children to her bosom at once.

I ignored the question. The boys looked suddenly important, while the girls glanced about as though they'd rather be off ahorseback.

"I don't believe this," she went on. "You may know something about birds, cows, and coyotes, but engineering?"

"It's all so simple. All I have to do is build a lens-shaped plug for the canyon 60 feet high, 250 feet thick at the base, with a drain pipe through the bottom."

"And a quarter of a mile long!" she wailed. "Look at these blueprints. What's this thing here called a chimney filter? And what on earth is 'freeboard'?" She left shaking her head, and right at a time when I myself was having moments of daytime terror.

Sensing my tensions, Coy got up from where she had been lying with the other coyotes, stretched, trotted over to me, and put both front feet against my belt buckle, grinning up at me. She had faith in us all. Trotting over to the dam site, she began to dig with her front paws. For a moment the dirt showered behind her, then ceased abruptly. She had uncovered a huge ledge of rock which would have to be blasted with dynamite. Later on I came to wish I'd quit right there.

Deep down I knew that all I was applying to the project was enthusiasm and the ability to plod along until the thing was done. Only I could judge my staying power. I wanted to have a lake so badly; to my mind, building one surely had to be part of the American Dream.

It wasn't my complete lack of engineering experience that bothered me as much as a lack of money. The ranch was already mortgaged to the hilt, and banks do not lend money on speculative projects such as lakes, especially when one can't project a cash return. It would take money to keep those ancient wrecks of machines in tires, parts, cables, oil, grease, and fuel. The plans and engineering had already used up what money I had managed to accumulate. Yet if I waited, with fuel costs going up daily, I'd never get the thing done. Rather than give up, I sold a few head of horses at an auction in town and purchased some diesel fuel.

In granting our permit, the state engineer had stipulated that the entire project be completed by the following winter. When spring runoff time came, logically, the dam and its spillway had to be completed. I carefully allocated my time and naively allowed three full days to the item marked "site preparation," which consisted merely of removing several thousand cubic yards of overburden from the dam site, peeling away all the rubble down to bedrock, and sweeping up any remaining dust with a broom.

Bedrock turned out to be some sixty feet down, and at the end of a month, I was still trying to find it. Every time I hoped I was through, the state engineer inspected and said *"Deeper!"* Lacking experience, I had failed to realize that some dams are like icebergs; there is more to them beneath the surface than above.

To make matters worse, the excavation kept filling

with mud, bogging the old tractor down past the limits of traction. It was exasperating to bulldoze one load of muck out of the hole, only to be passed by the previous load oozing its way back into the morass. The whole scene was a mess: mud everywhere, rocks everywhere, downed trees lying like jackstraws.

With the aid of my sons, who brought to the enterprise knowledge they could only have gained playing in their sandbox, we jury-rigged a system of cables to pull the old earth-movers through those La Brea-like tar pits and safely up the other side. We went so deep that from the top those giant machines began to look like children's toys, and the place began to resemble a giant open-pit copper mine.

At thirteen, Taylor drove the thirty-five-ton tractor, pushing each earth-mover through the borrow area until red dirt spilled over the sides. At fifteen, John drove a twenty-four-cubic-yard earth-mover, scraping up loads in the borrow area, hurtling down the valley to the dam to dump his load in a four-inch layer, and then highballing it back for another load. I drove the Cougar, and Dayton towed the vibratory sheepsfoot back and forth across the fill to firm the soil to a proper state of compaction. By the time the engineer finally approved the site as being prepared, we Hyde boys were a hardened bunch.

It had been ticklish work. If the soil was too dry, it refused to pack to standard and had to be sprinkled and redone. Any rocks larger than a tennis ball had to be picked out of the fill by hand and hurled down the face of the dam.

The earth-movers seemed to have a strange antipathy for each other, for they seldom both worked at the same time. There were snapped cables, shattered bearings, dead batteries, clogged filters, corroded points, overheated radiators, and flat tires. In one eight-second period of my life, I blew out two earth-mover tires that cost

3,600 dollars to replace, and that were not available from anyplace closer than Chicago.

If the day started well, by eleven o'clock I might be on my way to town sixty miles away for parts, new cables, or to pawn my teeth for money.

Once we had built the level of compacted dirt up to the original ground level, we laid in a 250-foot pipe and headgate as an emergency drain for the lake, cased it in concrete, then packed it carefully against leakage. Little by little, the outlines of our earthwork began to take form.

We were all a bit astonished when the project came to resemble the blueprints. "Just think," I said one night as the full moon rose to help us haul the last loads of the day. "When we've finished, an astronaut on the moon with a big telescope will actually be able to view our lake." The boys looked even more worried than usual.

Slowly, we became a team, although tensions built up between us, and we could hardly get through lunch without a shouting match. While I dumped my load on the dam, John would be roaring on up the valley in a cloud of red dust, headed toward Taylor, who, when he was not bulldozing away the overburden of pumice from the clay, sat waiting with Big Alice. Seconds after John had moved into position in the borrow area, raised his apron, and dropped his cutting blade to load, Taylor pulled back his throttle and hit the stinger protruding from the earthmover, pushing the machine forward. When the machine was full, John closed his apron, raised his carriage, and took off with a roar, hell-bent for the dam, passing my oncoming machine on the way.

There were days when we actually grinned in passing, elated by a feeling that we were really carrying the mail. But there were other days when we cursed the old machines. Cables snapped along with tempers, and parts broke along with our spirits. All too often we sat tense and depressed, waiting for a mechanic from town to get us running again. One more day wasted. One more day closer to the deadline in the fall.

In the meantime, at the ranch, Gerdi, Ginny, and Marsha were haying, riding, and otherwise carrying the burden of the ranch. They were under just as much pressure as we.

Each day we stumbled up before dawn into the same routine: Eat breakfast, fill the fuel barrels on the pick-up, grab the sack lunches, pile the coyote pups into the cab, and head toward the dam project.

The pups loved to ride in the cab, sitting on our laps, front feet on the dash, swaying with the curves, watching for chipmunks. To passing motorists seeing that forest of hair, it may have looked as though coyotes were driving the truck. Coming home, however, we made the coyotes ride in back, for the favorite pastime of a coyote pup is finding the rankest carcass available and having a good roll.

The grass and sagebrush meadow immediately above the dam had long been the fiefdom of a pair of wild coyotes who were less than pleased to have competition. Let Coy, Boy, Benji, and Nick stray from the immediate area where we were working, and the wild pair came bristling out of nowhere and sent the youngsters scattering hell-bent back to their human friends.

At first the pups treated the whole affair as a game. To date, nothing in their lives had prepared them for violence. They loped on ahead of the wild pair, playing with each other as they ran, wriggling and tail-wagging as though they knew the big coyotes would run only so far and no farther. But when the female of the pair caught up with Benji, bowled him off his feet, and cut a savage gash in his shoulder, the funning ceased and they scampered for their lives. From that moment on, whenever they saw the wild ones hunting, they rushed to the gaping mouth of the huge pipe beneath the dam and tumbled down the hole.

It became their lair, and the first few times they sought it out, they lay in the cool damp darkness of their sanctuary as machines rumbled overhead, and did not venture forth until they thought it safe. By peering up the bottom of the culvert, I could see their silhouettes in the dim light, as they stretched out comfortably in the man-made lair.

At first the wild coyotes respected the roar and thunder of the speeding earth-movers and kept their distance, but as the days passed they grew bolder and bolder until the pups could not venture out on the plain unless I walked with them. One day I topped the crest of the dam with my earth-mover just in time to see the pups drop into the pipe only one jump ahead of the attacking male.

I shut off my machine and watched from above. To my delight, in their relief at escaping, the pups tipped back their heads in the bowels of the pipe and began to howl. The more they howled, the greater the surge of echoes and echoes of echoes. What issued forth from that cavernous pipe had to be the most monstrous coyote howl ever heard in nature. Bristling with sudden fear, the wild coyotes skidded to a stop, whirled, and sped out for their very lives. Rushing pell-mell through the sagebrush, they made a beeline for the ridges to the north, and when I again caught a flash of them, they had gained the top of the ponderosa divide and were still running.

Once or twice after that, they tried to regain their territory, but the pups had discovered the use of the echo chamber, and on sight of the wildings, dropped into the pipe and began to sing. Even after the pups took over the flat as their exclusive hunting territory, they often retreated to their music room, where even the smallest sound was enough to set them off on a good howl.

Having their own hunting ground, the pups soon acquired enough proficiency that they had little need of prepared dog food. Day after day as we worked on the dam, the coyotes hunted mice and short-tailed ground squirrels, sometimes stalking them, but most often chasing them or digging them out. For a few moments they would make the earth fly, then they would check out the hole, thrusting their noses deep until only their eyes showed, drinking deep of the tunnel's inner smells. In the brush piles along the lake where I was clearing the bottom, they hunted rabbits and ground squirrels, tearing at the piles with their mouths, scattering the branches I had so carefully piled. With their new sophistication, they used teamwork. While one or two of them would blunder along out in the open, the others would quietly circle, waiting patiently up ahead for some foolish quarry to make a break for freedom.

But however much they prospered and grew handsome on their wild diet, they never missed begging for lunch. One by one, the big machines would cease their thunder, first the two scrapers, then the sheepsfoot, then lastly Big Alice, as John, Dayton, Taylor, and then I piled off our machines and sought noontime shade under the pines. Within moments the pups would materialize out of field or forest for the great social event of the day. Each occasion was the same. Nick and Benji would plunk down at stations ten feet away, with studied indifference, Boy would sit at our feet watching every bite of lunch we took, while Coy would crawl into our laps, covering our faces with kisses, turning her head sideways on occasion to let us know she would like us to scratch the inner cone of her ear. In a previous life Coy was perhaps a famous French courtesan, skilled at wheedling treasures from a king.

Beyond meat scraps, they loved cantaloupe and fresh peaches. The peaches they ate on the spot; the cantaloupe was good for a whole afternoon of games. One would seize the rind, mount a hillock, and defend the mouthful from all challengers, a game very close to one that Don Coyote had once played with me. Let the other pups lose interest, and down would come the pup with his prize, rushing past the other pups full tilt to restart the game.

Often as we sat, the Don came out of the woods and watched us, as though he remembered lunches of times past and was tempted to forget and forgive, but he never came far from the timber, as though he realized that his lack of a hind leg might make him vulnerable if he got too far from shelter.

Days passed into months. No longer was the dam just a dream on a blueprint, but an entity that had changed the face of the local landscape forever. Each of us had changed too in the process and would never be quite the same again.

The strain had gotten to us and transformed father and sons into gaunt-faced strangers, given to quick angers, sudden resentments, and week-long silences. We hated our machines as though they broke down just to vex us; we hated each other for many an imagined error or discourtesy, giving vent to rages close to hysteria. A day off to fish the Williamson would have done us all good, but there were only so many days left before the deadline, when snowfall would shut us down.

However outrageous we were to each other, the pups saw none of our anger. They were our one common joy; it was easy to talk to the pups even though we weren't speaking to humankind. Often we forgot out troubles to laugh together at their antics.

During those long summer days, the pups underwent a transformation. As with human children, there is a subtle point of growth when cute goes out the window. One day I looked at them and saw that they had lost the puppy look and were leggy and lean, able to race the wind. There was no mistaking them now for anything but coyotes.

As to behavior, Coy was still her sweet lovable self, while Boy was only a little less amiable. Nick and Benji seemed influenced by their stay with their parents after their eyes had opened, however short that period had been. They were more independent, more nervous, less adaptable to new situations.

At seven months, the four played well together, often to a point where joy welled up to such a crescendo that they would sit down on their haunches and have themselves a group howl, each caught up in the excitement of making a splendid sort of racket. Howling seemed not just a means of communication, but also a release of energy, grudges, and nervous tension. Once a song was started, they almost invariably continued it to the end, as though Nature had programmed them to sing the whole piece.

Playing in the midst of construction havoc, with all those machines rushing about, a dog would have been flattened by a tire. But natural timidity, alertness, and good sense kept the coyote pups from harm's way.

Little by little, Nick, Benji, and Boy began to show their independence. If in August, blended with the tan grasses of a dying summer, they were hard to observe, by October, with full winter coats, they were impossible. Where once they had waited at quitting time to hitch a ride in the pick-up, now they had to be coaxed to go along. Coyotes don't mind the way dogs do; when they don't want to do something, they don't do it unless tricked. I tried to counter the pups' reluctance by spending more time with them, stroking them, feeding them snacks, scratching their favorite itches. But clearly I needed them more than they needed me.

There was a new wildness in them. As they hunted out on the flat, they looked like any other coyotes, and until we called them and they actually came trotting in, we were never quite sure that we were calling our pets.

"They're going to leave us soon," Taylor remarked, a bit sadly.

I nodded. "All except Coy. If it came to a choice, I think she'd stay."

As if in answer, Coy trotted over to where I sat resting on a log, and leaned up against my knee.

24

NICK AND BENJI WERE the first to disappear. One evening, as we waited at the pick-up for the pups, Coy and Boy came trotting in to my call, but the other two failed to show up. I shouted their names, walking out on the cleared land to the edge of the ground squirrel towns where they loved to hunt. From a distant ridge I heard a coyote bark. I barked back an answer, but there was only silence.

Coy ambled across the field, sniffing the ground frequently, trying to pick up their scent, stopping now and then to stare across the landscape, ears pricked forward, looking first here, then there. From her indecision, it was obvious that she had no idea where they had gone.

At dark, we left without them, fearing the worst, but in the morning they were at the dam waiting for us, tongues hanging out, weary from the long hunt. They stayed around us most of that day, but that evening they again failed to show up. This time the wilderness claimed them, and though I saw Benji once more, never again did they ride back to the ranch house in the pick-up.

I was living alone now on the ranch during the week. The children were at Klamath Falls with their mother. The school year had begun, and the willow leaves had turned to gold. There was a snap of autumn in the air when Boy left us for the first time. Whether he found Nick and Benji in the hills and hunted with them I never knew, but at least he did not leave the area. Now and then as we ate our lunches, building a bonfire against the chill, we would hear a whine behind us as Boy trotted out of the woods to join us.

Tails wagging with quiet humor, Coy and Boy went through the rituals of greeting, licking each other's jaws, pawing at each other to induce play, and roughhousing. In the evening Boy might even choose to ride home with me for a big meal and a visit with the ranch dogs, but by morning you could read his restlessness to be off, and within an hour after his arrival back at the dam, he would be gone again. Clearly he was torn between two worlds, and he found the wild one more interesting.

One day when the dam was near completion and school was making inroads into my boys' time, Boy joined his sister and me for a sumptuous lunch of steak sandwiches, then ungratefully left us. I saw him trot to the edge of the timber and howl his locator howl. If, from afar, a friend answered his cry of loneliness, it was lost to me in the sounds the north wind makes as it sweeps through the pines.

I did not see him go for I was busy petting Coy. When next I looked across the flats where Boy had stood, I saw only a solitary sagebush dancing in the wind.

Without her littermates to play with, Coy became more attached than ever to her human friends, though there was still nothing in her demeanor that indicated she placed any stock whatsoever in pleasing us. However, we had become adept at tricking her into doing what we wanted her to do.

If she came when called, it was only because she wanted to come and at the moment happened to be doing nothing more interesting: Either she was curious, feared missing a tasty morsel, or hoped to go on a trip. It was always to please herself, but never, never to please me or any other member of my family. Let her get the idea you wanted to do something such as put a leash on her, and she would scamper away.

Housebreaking? Forget it. When I allowed her in the house, she loved it, but she was never a safe guest. I learned by experience to keep her in the smallest room possible with her own special blanket. This gave her a sense of being in her own lair. Coyotes go through their lives marking scent posts with urine and defecating where they please, but they do not soil their own nests. With Coy, if there were a rug you wanted to protect, you had best hide it, for if you scolded her and tried to drive her off, she would be back within moments sniffing about, puzzling over the rug as though wondering where she could best make her mess to drive you up the wall.

Afraid to lose her, I took her everywhere with me. The back of my station wagon became her special world. She rode with her ruff against my ear, muzzle against my cheek, watching each approaching driver as though searching for friends.

But the wild animal was still there. One day, as we drove south into California, a huge diesel rig roared around us to pass and, at that precise moment, blew a tire. Coy yelped in fright and completely lost control of her bladder. From that moment on she wet the car whenever there was the slightest hint of danger, and the air became so swiftly saturated with ammonia, there was little I could do but pull over and somehow clean up the mess.

Faced with solving the problem or leaving her home, I

built her a tiny kennel for the car and placed her security blanket on the bottom. She took over this den immediately, and from then on, as long as she could retreat to this safe house, she felt secure whatever the threat.

One household sound that never failed to alarm her was the collapse of a log in the fireplace or the stove. To her it somehow signified danger, and she would scamper away, trying frantically to tunnel out through the walls.

But there were other household sounds she liked. With Toby, the porcupine, she shared a fascination for the vacuum cleaner. My wife had only to start to vacuum the living room rug, and Toby would come crawling out from under the sofa and go bouncing along beside the machine, moving up to it bravely, then scampering away. Coy would crouch low in front, nose extended on her paws, her tail high, to await the machine. My wife had only to chase her with it, and away she would go racing from one room to the next, slipping and sliding as she crossed the bare pine flooring between rugs, leaping into the seats of easy chairs and vaulting over the backs, figure-eighting about the dining room, making shambles of the beds in the bedrooms as she cavorted about, but rushing back to challenge the machine and coax it to chase her off again.

Jack had been terribly one-man about his affections, and the other pups had kept their distance until they judged a person a friend, but Coy knew no strangers. Old friend and newcomer alike got the same treatment. Like Browning's Last Duchess, she "smiled at all she looked on, and her looks went everywhere." She greeted everyone with the same effusive greeting, leaping high to touch a person's jaw with her wet, black nose.

I arrived at the ranch one evening to find a Jehovah's Witness sitting reading his tracts on the lower branches of a lodgepole pine, while Coy lay stretched out below, cocking her head now and then as though wondering why her new friend did not come down to play.

To Coy, strange children were only an extension of my family. If they had not been taught fear by their parents, children accepted her kisses and hugged back as though she were an affable shaggy puppy. They wandered back to their parents not a bit the worse for wear unless their parents minded them wearing the distinctive perfume of carrion.

At times in my life when I felt a craving to be alone or had to work cattle on the ranch where I just couldn't have her tagging along, I had to lock her up to keep her from following behind my horse. By now Coy was adept at reading body language; it seemed to me she could read my mind and sense my evil intentions almost before I thought them. Perhaps she read, smelled, or sensed tensions about me of which I was unaware.

To catch her, I had to put the undesirable act in the furthermost, cobwebby corner of my mind, call her to me, play with her, or sit scratching the inner cone of her ear until she had completely forgotten that I had summoned her. Then, even though others waited for me, muttering under their breaths for me to hurry, it was vital that I take her on a fast fun romp around the area. Only then did I dare tie her up with any certainty that I would ever be able to catch her again. If she ever associated that leash with anything but fun or food, the sight of it would send her running.

Coy's sense of hearing was astonishing. She could be busy playing outside in the garden when I, sitting inside the ranch house, rattled a piece of cellophane on a doggy treat, and there she was, quite suddenly, front feet on the sill, peering at me through door or window. I had to feed her right then, or she would stay mad at me for hours.

If I treated Coy as I treated my domestic dogs, she was quick to straighten me out, to remind me by her behavior that, much as she liked me, she was still a wild animal. Like her father, Jack, and the rest of my coyote friends, she was easily piqued, as quick to do a turnabout as a

March wind. She used ostracism as a tool, as though somehow she sensed my need to be accepted.

But I too had my tricks. I was learning the ways of the coyote. She too was vulnerable. However stormy her mood, I had only to sit with her and howl, as I used to with Don Coyote, and the mood was forgotten in her excitement at having a friend with whom to sing. She was as unable to resist a good howl as a dog an attack of epilepsy. And once the howling was triggered, she seemed programmed to go through the whole chorus, head tilted back until it seemed it would fall off backward. When our last wavering duet would cease, so would her snit. Mellow with forgiveness, she would rush into my arms, wriggling, smiling, licking my jaw as though the mad had never been.

Coy loved playing with the dogs, but her great reserves of nervous energy were too much for them. She kept on playing interminably, while they, after a few hard tussles, lost interest and tried to retreat to the shade. Creeping up on them, she played at killing, pouncing on them with gaping mouth, chewing on their ruffs, ears, legs, tails, anything that presented a target. Fortunately for the dogs, her grip was of rubber rather than iron.

She was also an actress. Let a dog lose patience and turn on her, and she fell on her side, belly up, smiling her coyote smile, licking its jaw in supplication, pawing with love, lifting her leg to show her parts as though to say, "Hey, no fair, I'm a girl!" It was all playacting, letting the aggressor think he was superior. When the dog fell for her act and let her up, she was back moments later, bullyragging as before.

If a strange dog rushed her, however, she could become frighteningly hostile, her posturing resembling that of a wild stallion challenged by a rival. A stallion drops his head, ears laid close to his neck, eyes snapping with rage, moving forward serpent-like, in a weaving, shuffling dance, purposefully awkward, jolting the ground

with his hooves not quite on target until, suddenly, like a
flash, the head comes up, the mouth gapes, exposing an
array of teeth ready for the bite, and squealing, he ex-
plodes into an attack.

So too would Coy posture, head dropped to protect
her throat, ears flattened back, face split until it became
all mouth and teeth, body laterally compressed, back hair
and ruff bristling, making her seem twice her size. Stiff-
legged, head down, she would approach the enemy. If
her adversary wagged his tail and stood his ground, the
encounter ended as her glance seemed to stray to other
objects and she tried to save face by becoming preoccupied with sniffing a nearby bush or tree.

But let an interloper turn to flee, and Coy was after
him in a flash, never ceasing her attack until the interloper
was driven from the territory. Victorious then, she staggered back, glancing up at me to read my reaction.
Tongue hanging out as though she had forgotten to
breathe, she usually plunked herself down at my feet.

By deserting me, the other pups had left a void in my
life, but Coy filled it admirably. Some innate wisdom
within me, however, guarded my innermost heart.
Friendships with wildings are too soon over; inevitably
the days are numbered.

25

IN THE FALL THE boys had gone back to school with a feeling of fulfillment.

"Guess what, Dad," John had told me over the phone from town. "Today in English we had to write the true story of what we did with our summer vacation. I gave 'em ten pages. All the teacher said when she read it was I should write novels."

"And, Dad," Taylor had chimed in from the upstairs phone, "my teacher couldn't get over our building a big lake just for wildlife. She didn't even know what 'habitat' meant or that you've got to have predators like coyotes around to keep wildlife healthy. What a dumber!"

But it was not until autumn had passed and the last load of dirt was tamped into the dam that the boys, with a sigh of relief, could go back to being kids again. As they drove off to town, I faced the gnawing realization that tomorrow the state engineer would arrive with his crew and either pass on the dam or tell me to tear it down. It could go either way.

On the day of my deadline, right on schedule, the survey crew piled out of a state vehicle parked on the dam. There they were—the lean, the rotund, the windburned, the pale—setting up their transits and test equipment, prodding the earth, digging, taking core samples, muttering under their breath, and holding the future of my dreams in their hands. The sun chose that moment to go under a cloud, and I feared that it was an ominous sign. As they worked, sighting and measuring, they shook their heads, made notes in little notebooks, and conferred seriously with one another.

To keep out of their way, I went off into the woods with Coy and back to the dam site only when the big chief engineer himself arrived to hold council with his braves.

He glanced at me without even a smile. Ever since he'd found out I had a degree only in English, he'd been hard on me. I, who defined a grade stake as "a wooden stick covered with hieroglyphics driven into the ground in the immediate path of an earth-mover tire," had found his lectures, delivered under a hot summer sun to the roar of an idling earth-mover, boring and unintelligible.

By his endless repetition, I had come to understand that "freeboard" is the distance from the high-water line to the top of a dam; that "camber" is the arch of a dam toward the center to allow for settling; that a "chimney filter" is the internal wall of crushed rock standing tall within a dam to drain any possible seepage out of the structure and carry it downstream through a T of perforated pipe; and that "slope" is the angle of rise of a dam from the floor of the valley. I also learned that smiles are not part of an engineer's daily repertoire.

Now, as he approached where I sat on a huge boulder awaiting my doom, he gave me as much of a smile as a state engineer is allowed to give and admitted that a professional dam builder, because of the costs involved, would not have built such a good edifice.

I sat there staring at him dumbly. Why hadn't he given me a few kind words like that during the summer when I needed them? I hugged Coy to my chest, unable to give the engineer more than a stupid grin. The dam had passed! It was now ready for filling!

I shook hands weakly and, when they left, wandered off through the trees, with Coy trotting at my heels. Somehow we had managed to build a dam for a lake where none had existed for many thousands of years and had addressed our feelings for wildlife by teaming up with Mother Nature to create a habitat.

Maybe the lake would be more than just a body of water on a ranch in Oregon. Maybe, as an oasis in that arid pine forest, it would become such a wildlife spectacular that I could use it to inspire other private landowners to help wildlife on their lands. The hair stood up on my arms and a shiver went down my spine as the idea germinated and broke ground. The concept of private land coming to the fore as the wildlife refuges of the future boomed within me, but as usual, I had to stand at the sidelines and wait. Could I really count on a lake? Maybe there was a hole in the ground out there that would suck away the water. Right now I was looking at nothing but the bare bones of a man's dream. I would have to wait out the long winter until snow melt time, when water would come flooding down the canyons.

After the inspection team had gone, I rushed back to the ranch and phoned my family, telling them the good news, wishing that we could all go out for dinner to celebrate, but accepting the reality that they were living in town sixty miles away, and I had cattle trucks to load that night or our yearlings would never get to market.

I celebrated instead with Coy, pouring her a puddle of beer on the kitchen linoleum. We were both having a pretty good howl together when the flash of running lights through the pine trees told me that the trucks had arrived.

During the next few days, I finished some touch-up work on the lake basin, piling brush with Big Alice. Each day as I moved the old tractor, I noticed fresh scrapes beneath it where Don Coyote had spent the night. Now and then I saw him out on the ground squirrel towns or chasing chipmunks among the pines, but he ignored Coy and me as though we were invisible.

Once the cattle were shipped, one more task faced me before winter—disposing of the old buckaroo line cabin at the far end of the flat. Sometimes as I inspected it, I thought that fire would be the most fitting end to the cabin, pulling down a swift curtain on memories.

Never painted, it stood sun-warped and wind-battered. The windows had been vandalized, the sashes and glass stolen or bird-busted. The blistered door sagged on one lone, tortured, rusted hinge. Home now to porcupines, pack rats, spiders, flies, flickers, bluebirds, starlings, bats, and swallows, it was totally unfit ever again for human habitation.

Cowboys who had worked for my uncle had used it Octobers and Novembers during the fall gather, as they hunted bovine survivors along the Sprague River Valley, where many a man brought up his kids on stray Bar Y beef. In the smoky wood of spidered rafters, names were carved: Ash Morrow was there, and Homer Smith, Ern Morgan, Slim Fields, Hank Hall, and Roy Wilson—all gone now, riding the pastures of beyond.

For years we had kept a cabinet of food, a stove, and a supply of firewood for any winter-chilled stranger who might need emergency shelter on the long, empty road from the ranch to town. This was common courtesy in the land.

But values changed. There came a time when thieves stole the woodpile rather than cut their own logs; when vandals shot bullet holes in the canned goods, broke legs off the big oak table for firewood, and finally stole the big cast-iron cookstove itself. Eventually, the shingles gave way to the prying wind so that the roof leaked and water pooled in bat droppings to fade the bright blue flowers on the linoleum.

One could only enter the sad, dank shack with a yearning for the sunshine that had once made a sundial on the floor and for the laughter and cowboy stories of times past. Now, with the feeling of decay, came the certainty that the past was gone and there was no bringing it back.

Trying to decide the building's fate, I sat on the porch in the thin sunshine, leaning back against the weathered boards, with Coy crowding my knees, as I looked off across the frost-burned meadows. I saw the gully which, one winter long past, a pranking wind had blown rim full of snow, and when I had tried to ride across it to join a couple of Indians on the other side, my big paint horse had stumbled, spooked, and bucked me off head first into the drifts. I can still hear their hooted laughter.

I saw the thicket, as yet uncleared, where a stampeding colt named Cyano broke the leg of a cowboy. Clad in a navy blue peacoat against the cold, the cowboy started the long crawl back to the cabin, but the weak-eyed old caretaker, Mac MacAlester, who saw him approaching, mistook him for a bear and emptied his rifle at him. Mac was hunting in his pockets for more ammunition when he finally detected cuss words above the wind.

I remembered sitting at Mac's table with Whiskey Jim McGilvray when both of us were boys, trying not to hurt old Mac's feelings by laughing when he lost his false teeth in the gray, grimy dishwater, then slapped them, dripping, back into place.

Though, looking back, the wise thing for me to have done at that moment would have been to touch the place off, something, maybe a ghost of a cowboy past, kept blowing out the match. So instead, I jacked up the shack, pushed two big ponderosa pine logs beneath it, and sledded it behind Big Alice across the withered grasses. It was like putting a pistol to the temple of a favorite old saddle horse, then relenting and giving him one more summer, although shooting him then and there might have been more merciful.

Twice on the trip across the flats, the cabin lurched on uneven ground and fell off the runners. Twice I jacked it up, replaced the skids, and took off again. I took it to a point on the east end of the dam, and despite my having thought to destroy it eventually, it looked as though it belonged there, nestled comfortably amongst the pines.

Jacking it up high, I settled it on huge flat boulders from the dam excavation, then contemplated my work. Built at a time when lumber was cheap and even a buckaroo camp could be built to last, it now stood solid as Gibraltar. There was even a faded charm about the weathered siding, and the roof could easily be reshingled against winter storms. As to windows, I happened to know of some stored in an old shed that might just fit.

I felt a strange excitement. Down deep, I was tired of trying to heat the big ranch house at Yamsi while Gerdi and the children were in town for school. I wanted a place like the old Hyde camp in northern Michigan where I could toast my aching bones before a roaring fire in a stove and where Coy could spend the nights indoors without the trauma of coyote hairs on the furniture.

Yamsi was always so neat and clean. My wife always left it as though she feared she might die on the road and some other woman might come in and pass judgment on her. I saw the old shack as someplace where I could relax with my feet up, drop my socks on the floor if I damned pleased, and leave them there until the next load of wash.

For years I had been a weekend father, and now my ways had reverted to being as solitary as they had been during my boyhood. Much as I loved my family, I found an increasing contentment in being alone, walled away

from others of my kind by miles of forest, with only Coy
for company.

During the week there was wood to cut and hay to
haul into the barns and sheds. There were also hooves to
trim on the big draft horses that pulled the feed wagons
through winter drifts, calves to wean, and cattle to vacci-
nate and spray against lice and grubs. Each night I fell
into bed in exhaustion.

Although the cattle came in easily off the ranges, there
were days when I spent every waking moment shipping
cattle or getting my herd ready for winter. Moments
after the family arrived from town the children were
ahorseback, helping me with tasks I couldn't handle
alone.

Little by little, however, I stole time away from the
ranch to fix up my camp. I put on new shingles so that
the rain no longer puddled on the floor, restored the door
to its hinges, and replaced the windows, to the disappointment of pack rats, mice, and pine squirrels, who still
considered the place very much their own. Then came
bright new linoleum on the floors so that the cabin could
be swept clean, and a barrel stove so that the damp of
decades could be driven away.

In the Yamsi dump I found the big white kitchen sink
that had once graced the ranch house, and I installed it
not with running water, but with a drain pipe out the
wall. For light, I restored some old Coleman lamps. With
the addition of a wooden bed and the old buckaroo chair,
the place became almost habitable.

By day I worked busily, shoveling out, cleaning, and
scrubbing, but by night the other inhabitants worked just
as hard turning the place to a shambles. Each morning, I
arrived to find nests of pine needles and trash big as
bushel baskets on sink and table, as the pack rats found
their way around the holes plugged with steel wool and
tin can lids.

But the animals had reckoned without Coy, who swept into the cabin like a prairie storm: The cobwebs she carried away on her coat; the animals she kept treed in the rafters until they lost patience and departed via a hole made by a flicker.

One day I waited for the exodus, then dashed outside and nailed a piece of tin over the hole, assuming that this time of year any self-respecting flicker would be wintering in southern climes. But this one had apparently been using the shack for winter headquarters, for I had hardly blocked the hole, when the flicker appeared out of a neighboring pine and proceeded to drum a protest on the obstruction.

Only the mice persisted, slipping through the tiniest holes. One by one I closed them using a caulking gun on the cracks until the last remaining mouse couldn't find its way outside. I named her Mrs. Flannigan, and resolving to coexist, put out a saucer of milk for her on a shelf Coy couldn't reach.

It was not a good idea. Mrs. Flannigan acquired the knack of letting wild mice in the front door and gave wild parties. It was when I spent my first night in the cabin that I discovered that the mice were still very much in command.

Shortly after I turned off the lamps, bedlam started. In a wild leap toward the rafters after a mouse, Coy managed to knock a fifth of Jack Daniels down between the studs where I couldn't retrieve it without knocking out a wall. The odor of whiskey pervaded the room, and a game of lick-the-cork became the order of the night.

Neither Coy nor I got much rest. She crawled up on the bed beside me and lay awake watching. Every time a drunken mouse staggered across the floor, Coy catapulted off the bed and sailed across the linoleum like a hockey puck on new ice, ending up mouseless in a heap against the far wall.

The night was filled with the sound of pack rats lawn

bowling on the roof, the dancing of mice indoors, and the gloom of horned owls in the surrounding pines. Coy and I got our best sleep when daylight came and the night folk flew or staggered away.

I was so proud of my new retreat that I invited the family for a Saturday night dinner. Thawing out a big beef roast, I put it in a covered pan and gently roasted it on the barrel stove all day long. Delicious fragrances filled the room. Mrs. Flannigan came out to look, and even Coy was beside herself, stalking back and forth before her erstwhile enemy, the stove, working her black nostrils as though wondering from whence those heavenly odors came.

While the family sat stiffly around the room on apple boxes, glancing about as though remembering the shack's previous condition and feeling uncomfortable just being there, I was rattling on about Mrs. Flannigan and her friends. All of a sudden the smell of smoke made me realize I'd burned the peas.

"That's quite all right," I said. "We've still got the meat and potatoes." I opened the pan to let a reasuring fragrance fill the room.

"What are these, Dad?" Marsha asked, spearing the charred remnant of a potato with a fork.

"Well, there's always the meat," I said, setting it on the drainboard of the sink to cool.

Everyone sat around, probably thinking I couldn't possibly do any more damage to the meal since the roast was done to a T. I was sharpening a carving knife on my boot, chortling happily to Taylor about our experiences building the dam, when Ginny came up with an expletive unbefitting of a lady. I looked up just in time to see Coy dash out the door packing the whole roast in her jaws. All that was left in the cabin for seven hungry people was one can of Van Camp's baked beans.

ONE MORNING I AWOKE to a different world. The first snows of winter had turned the anemic, faded grass to a frozen deathscape of white. From my bedroom window at Calimus I watched the pines flinging restless silhouettes and heard great thumps, like far-off howitzers, as bending branches dumped loose clods of snow, darkening the panes with the dust of ice crystals.

Bounding out the cabin door, Coy skidded to a stop, amazed and frightened by the transformation. It was her first snow. Rushing back into the cabin, she hid behind my legs, peering up at me to see whether or not I was afraid. To give her confidence, I took a few barefoot steps out into the snow. She came out like a racking show-horse, tucking her feet high under her belly, still unsure.

When finally she relaxed, she rolled over and over and the icy fluff dry-cleaned her coat. She finished on her back, four feet waving in the air as she glanced at me and sneezed in the bright glare. Then, racing about the woods like a dog just out of a bath, she moved farther and farther away with each circle and soon was gone.

I leaped for the comfort of the cabin, drying my cold feet by the fire, then put on my boots and coat. Afraid that some harm might come to her, exposed as she was against the white of the snow, I waded about looking for her. She heard me call but ignored me.

Sometimes as I called she would creep up behind me and sit, regarding me quietly as I roared her name, listening bemused to my echoes, letting me be a fool in my parental worry. Now I whirled suddenly in the snow, hoping to catch her there, but there were only the dinosaur tracks my boots made in the powder, monstrous beside her dainty paw prints. No use tracking her; in the snow she could make ten yards to my one.

From the ridge to the north, I heard the hue and cry of hounds. Perhaps they belonged to government predator control men who were spending my tax dollars, taking advantage of the fresh snow to track down prey, hunting around my land not because of complaints, but because there were coyotes there. I thought of Benji, Nick, and Boy, as well as Don Coyote, hoping they were off in the back country away from roads.

I had hardly entered the cabin when a clacking roar sounded through the forest. I rushed back out just as a helicopter swept over the dam. In the right door of the plane I could see a man with a red cap, ear flaps down, hands gloved, holding a shotgun in readiness.

Unaware of the spectacle I was making, I ran after the helicopter, shaking my fist and shouting curses. For a few minutes they patrolled my forest, trying to make sense of the maze of tracks Coy had left in the snow. Soon they lost them in thickets west of the dam and swept eastward off my property over the shoulder of Calimus Butte, headed perhaps for the Sycan country. A few hours later they passed over the cabin again, with a string of gaunt, frozen coyotes dangling from a wire. It is up to the children of such men to be ashamed of their fathers.

I was still stewing about Coy when she charged me from ambush around the corner of the cabin and leaped so quickly into my arms that she knocked me backward into the snow. I lay laughing in relief, letting her lick the snowflakes from my face.

A few minutes by the fire and she was restless to be off again. I ignored her, throwing her into a snit. She retreated into the bedroom, where she worked at sound effects, tearing up a magazine and chewing on what turned out to be a button on my quilt. When I ignored her still, she carried a boot up on the bed in her mouth, then dropped it over the side. I tried to go on typing a letter, but she started scratching on the bed, and by that time I was writing four words and erasing three. Giving in, I opened the cabin door for her. She whined a quick invitation to join her as she swept out the door, but vanished before I could get on my coat and go with her.

It was past midnight when I heard her scratch to come in. She beat me to the bed, then defended her rights to it with a growl, but I ignored her and climbed in. It did not concern her in the least that I wanted to sleep. She leaped from the far side of the bed, down the crack between bed and wall, scampered and clawed her way under the bed, circled around the kitchen, then leaped back on the covers, landing on me with a stiff-legged pounce.

When I gave in and arose at last to play with her, she lay in ambush under the kitchen table, biting at my shins as I passed. Again and again she hid, dropping to her belly, nose outstretched as though to blend with the flowers on the linoleum. It was a game she had played a thousand times with her littermates, and now I was the mock prey. I was tired from a long day, but I went along with it because I valued anything that seemed to strengthen our relationship.

She lost interest in me suddenly for something outside

the cabin, sniffing back and forth along the wall, nose pressed to the mopboard. I let her out and next morning found fresh blood and the tiny plume of a pack rat's tail on the snow.

At Yamsi life was under the regime of winter. I had hired two men to feed the cattle, and they rose at daylight, cooked a quick breakfast, harnessed the big bay team of draft horses, then hitched them to a wagon or sled loaded the evening before with baled hay.

The hooves of the cattle made squeaky, crunchy music on the crisp snow as they milled about, sampling each bale as it was fed over the side. Other than feeding and cutting wood, there was little else the men could do in this weather.

I was a fifth wheel and spent more and more time living in the cabin, with or without Coy, for she often abandoned me for two or three days at a time.

Perhaps she might not have come back to me at all had not the solitariness gotten to me, and running out of patience, I saw no reason why she should not be there to take care of me. It was usually a simple matter of wading through the snow to a point in front of the cabin and giving my most enticing "Come home; all is forgiven" howl. Much as I sounded like a creaking gate, I was at least distinctive, and Coy knew my howl. Often as not, back would come her answering call, followed by other coyotes scattered through the hills who sounded as though they were laughing at me.

I came to know each family's music from the others. Coy's voice was distinctively shrill, as though living in the cabin had dried out her vocal cords. I would pick it up first in one part of the forest, then another, as though she visited all the families and had no consistent range.

When I called her after an absence, her voice had a joyous ring, as though saying, "Thank God! I thought you'd forgotten me!" I had only to raise my voice to the sky, then bide my time, and eventually she would come running across the dry lake basin, plowing snow with her chest. Invariably she was bedraggled and exhausted, but with a stomach distended as though she had dined at many a coyote banquet. Sometimes, as though in joy at seeing me, she regurgitated partially digested tidbits in my lap.

Mice and deer carrion made up the bulk of her offerings, but occasionally I found the ridiculously long tail of a kangaroo mouse. Rather than hurt her feelings at such times of reunion, I would dig a hole in the snow with my hands and bury the offering as though letting it ripen for a later meal.

In January a deep fluff of powder snow cut us off from civilization and made the going tough even for Coy. Like a bridegroom, I enjoyed the simple life, isolated with my friend. I awoke each morning to her persistent badgering to be let out to explore the world. First I heard her toenails clacking on the linoleum as she paced the floor, then scratching at the door, and finally, if that did not get me out of bed, she would bound up on the covers and begin to chew on one of my ears.

Once the door was opened, she sailed over the drifted doorsill, hit the far side of the porch, and catapulted into piles of fluff. After she had paused to burn a daffodil in the snow, she had herself a good roll, first on one side, then another, then circled the area to see what visitors the night had brought. Sometimes we found coyote tracks that traveled a beaten path around the cabin. Whether they had been made by Benji, Nick, Boy, or even Don Coyote I had no way of telling.

As a human, I admit to certain species talents, but sometimes Coy made me feel like a clod. Like any coyote, she seemed to possess a vision of what went on under the snow. Her ears, funneled forward, sought out the tiniest sounds, at which she would plunge her muzzle into the drifts until only her eyes showed. Often she came up with a struggling vole or, on occasion, an interesting pine knot, giving each the same kill treatment until she became bored and either ate the animal or lay and chewed the wood to splinters.

The snow froze to her coat in ice balls which rattled as she walked. The clicking ice hampered her travel and irked her until she plunked herself down in the snow and tried chewing them off. Soon, however, she determined that it was far easier to beg her way into the cabin and stretch out near the stove to let them melt. While I was proud of her ability to solve this simple problem, the process always left a slop of water pooled on the linoleum.

Often, kitten-like, she played games with the melting ice balls, batting them with her forepaws until they skidded across the floor. Delighted, she stalked them, leaping to the kill, a process that sent them skidding again.

At times she took the ice balls gently in her mouth and deposited them in the middle of the bed, curling up beside them as though they were puppies. Tired of wet bedding, I finally locked her out of the bedroom or took to currying her off before I let her in the house.

When Coy scattered my piles of precious pitch kindling in the snow or shredded my down sleeping bag to make a nest, our relationship wore pretty thin. But just as I would be feeling a need for my own kind, there would be a hunt alone with Coy on a wintry night, on the snowy desolation that would someday be a lake, watching as the night world became almost light again from myriad stars. Coyotes would howl from the surrounding hills; Coy and I would answer, singing our duet, and the joy of being would be almost too much to bear.

One night I paused with Coy in the middle of the flat and listened to a flight of whistling swans passing overhead. Wings creaking, they made a wide circle as though searching for open water, and finding none, went on. From the dark pines along the shore, a horned owl shrieked at us, and Coy, perhaps feeling vulnerable in the middle of that expanse, stayed close by, walking in my moon shadow and hovering there until we reached the safety of the cabin.

That winter we became wondrous night stalkers, Coy and I. How often we watched as a great yellow orb of a moon rose over the Calimus ridge, diminishing in size as it left the magnifying atmosphere, hardening from soft butter to pale ice, but scattering on the snow between us a path of glistening pearls. With the rising of the moon came a few wakening breezes, stirring the pine tops from their slumber.

Coy liked to hunt where the sweeping winds had laid bare the grasses, and when she would freeze, ears rigid, only her eyes moving, I would honor her point. My stupid vestigial nostrils flared to catch a scent too faint; my ears strained to hear sounds not of my register. When she pounced easily and caught a vole I hadn't even seen, I felt hopelessly inadequate. But there it was—proof!—struggling its last fight, squeaking angrily as it fought for life; air, seed, and dew turned into protein and about to become part of a coyote.

One night in February Coy killed a brush rabbit caught out in a patch of squaw current bushes away from the timber. As I sat waiting for her to finish her leisurely meal, I shivered, but as much with excitement as with cold. It was a time when coyote estrus was at its peak, and noisy and excitable, the animals howled with the least excuse.

The jet plane to Portland and Seattle blinked in the sky, filled with sleepy people lost in their own thoughts, unaware that beneath them a man lived in a shack with a coyote or that their very passage overhead sent the forest denizens into a chorus of yapping.

Other sounds too—a flight of Canada geese, tired of Sacramento Valley rice fields, yearning for the Slave and the Athabaska or the Yukon Delta; a tree exploding with frost; the slamming of the cabin door in the wind—set off a ripple of coyote hysterics, loner to loner, family to family, rippling in ever-widening circles.

Tomorrow the snow would be riddled with tracks of restless females, leading retinues of amorous males, circling and backtracking as though to leave no tree undrenched with urine or patch of snow untracked.

As I stood in my open doorway listening to the chill night, I heard growls across the snowy wasteland, even savage combat, as a female told some prancing, eager swain that he was yet too early. When the time is right in estrus, there are no unwilling females.

I left the comfort of the cabin behind and moved to another listening point, where Coy came and sat beside me. Now and again, trembling with excitement, she licked my ear, making it burn from the cold. From the north rim, a solitary coyote barked and howled. So familiar that howl!

Minutes later I heard that same coyote again at the edge of the timber and saw it trotting in the clearing, sharply etched against the snow. There was no mistaking the animal. No tail and traveling on three legs. Don Coyote himself!

Coy saw him too and trotted forward. Midway she stopped, looked back at me, and whined, wanting me to follow.

"Get on with it, Coy," I called. I was trying to be fair and let her go off to enjoy herself, but I was worried too, since the woods were full of females that would not appreciate an unattached female, even a youngster like Coy, too young to come into heat.

At the sound of my voice, she came running back to me, touching my ear briefly with her nose. Lost in thought, I did not see her go. From the forest north of the point, I heard Don Coyote again and answers from other lonely hearts, each in his turn.

The comforting light from the Coleman lamp left burning in the cabin beckoned me home. For a time I sat

at my crude desk, going through the motions of writing
a journal. Mrs. Flannigan gnawed at a new access hole to
the cupboard. I threw a boot against the wall, but she
knew my blusters to be harmless and, after only a moment's silence, went on with her carpentry.

Two days later I was distracted from writing letters by
whines. I went outside and saw a bedraggled but happy
coyote dragging a huge snowshoe rabbit toward the
cabin. She seemed pleased when I stopped in the snow to
take it from her. It would make a stew we could both
enjoy.

Coy had a nick in one ear and a raw, diagonal fang
slash across one shoulder; from her hip she had lost a
patch of fur as big as a silver dollar; she limped painfully
where she had been bitten on her left front leg. But she
seemed happy. Most likely she had wandered into the domain of a jealous female and had been forcibly ejected. It
was a week before she left my side again.

27

COY THUMPED HER TAIL against my knee as we sat
and watched the first blue-gray trickle of melted snow
water flow, like lead into a bullet mold, down into the
lake basin. On its surface it bore a micro-fleet of tiny
boats: weathered bud cups from the pines, wing cases of
pine beetles, fairy fishhooks from a cone, a pine nut hollowed out into a dugout canoe by some feasting grub,
fragments of pine needles fractured perhaps by a deer's
hoof.

It was not an auspicious start. The moisture content of
the snows had been light, and the storms, which usually
bury the surrounding mountains under a pack of snow,
had all taken jet streams east and dumped on New York
folk who probably hated them. Without cover, the manzanita bushes on the higher elevations froze and turned
brown, as though Nature had wanted to extinguish the
species.

The thirsty soil on the stream bed had been soaking up water like a blotter for a few days when one day Coy was skipping blithely across a carpet of pine needles along a favorite trail, and, to her astonishment, fell through into water. Dripping and embarrassed, she emerged on the far side of the pool, shook off the needles clinging like quills to her coat, and gave me a hard, suspicious look to determine whether or not I'd been involved. I made the mistake of laughing at her, and she would have nothing to do with me for the rest of the day.

That night a warm rain fell, melting the skimpy snows on high and sending a gushing torrent down into the basin. It coursed down over the dry lake bed, piled up against the dam, and began to back and fill. First a puddle, then a pond, then a small lake. It wasn't the body of water I could expect someday, but it was exciting.

One day as I sat at the listening point in front of the cabin, wishing for a cloudburst, I heard a mournful whoop from the gray, scudding clouds. For a moment the mists opened as for an angel and the sun's rays made a golden stairway down to the land. Down that stairwell glided five great white whistling swans. Circling the lake they whooped in apparent disbelief that such a body of water could exist there.

For several minutes they circled suspiciously, brilliant white against the dark of the pines. This was the payoff. It was as though Nature herself had decided to acknowledge our desperate, hard work; I missed having my family there to share it with me.

At last the swans completed their inspection. I held my breath as their great white wings ceased to oar and froze outstretched, flaps lowered, as they began their long glide downward. I saw the mirror shattered, great splashes of liquid mercury as they beat backward to slow their speed. Landed now, they regrouped, whooping together like excited children; drinking, bathing, preening, tipping tails as they fed off the bottom; then resting as the lake calmed and the mirror mended itself. Double images in the calm, they were ten swans now instead of five, half upside down: two adults and three immatures, the young off-white, as though they had used the wrong detergent.

Caught up by my excitement, Coy wagged her tail and waded chin deep into the icy water toward the newcomers. In their curiosity the swans swam lazily toward her. At their approach Coy became suddenly unsure and found things to do ashore, stopping now and then to make sure the big birds were not coming after her. She saved a pelt load of water for me, drenching me as she shook.

That night I dreamed that the swans were so pleased with my lake that they forsook their arctic homes to stay and raise their broods. I awoke from the dream with a sharp pain in my chest and trumpets sounding in my ears. For one wild moment I thought the end had come. But the pain was only Coy wanting out, standing with all four feet on my chest, and the trumpet was a brassy farewell chorus of swans.

Coy leaped to the floor while I followed, barefooted, muttering obscenities. The great swans were in full voice as they prepared to leave the lake. Soon, in the darkness, I heard wings beating heavily, the creaking of wing tendons as the swans became airborne. They lifted over the cabin roof so close it seemed I could have leaped to touch them. Coy tried. She soared high after them but landed flat on her back.

For a time I watched them in the moonlight as they circled a farewell and got their bearings. When only silence remained, I felt a wet nose touch my hand, as though Coy sensed my sadness and was trying to comfort me.

I needn't have mourned their passing. The next morning there were more swans, and rafts of ducks and geese. Like an irresistible magnet in this arid land, the lake drew waterfowl out of the sky. Like a second magnet, Coy

drew them in flocks to follow her just offshore. It was a
phenomenon as old as waterfowl themselves. Ancient
man must have marveled at the sight and gone back to
his cave to make up fables.

Perhaps it is curiosity on the part of waterfowl, perhaps
the fascination so many prey creatures seem to have with
their predators. It is like a sick man looking through a
microscope at the bug that is killing him. When Coy
trotted along the shore, buffleheads, goldeneyes, scaups,
ring-necked ducks, pintails, wigeons, wood ducks, teals,
as well as Canada geese and swans rafted along behind
her, towed by their fascination as by a string.

At first, like any other youngster, Coy gave chase, but
they retreated just out of reach into deeper water and
shamed her. From then on, she pretended to ignore
them, hunting mice with great concentration along the
beach, as though the waterfowl didn't exist. In reality,
she was acting. She was only too aware of their presence,
patiently waiting until one grew too brash, too careless,
and ventured too close.

The lake grew from ten acres to fifty, to a hundred.
Now and then I would see a small aircraft flying north or
south do a confused turnabout, as the pilot noticed a
body of water not on his maps, then retraced his course
back to some familiar landmark. Then back he would
come, straight as an arrow, crowding his engine, impa-
tiently trying to make up for lost time.

Now and then a plane would fly low for a closer look,
but the clouds of nervous waterfowl rising from the lake
had a way of keeping pilots at a distance. Without knowing it, the ducks and geese were protecting their airspace.

The waterfowl did less well against the bald eagles,
who, attracted by the concentrations of birdlife, adopted
the lake as a new territory. The big predator-scavengers,
our national birds, picked out high perches on dead snags
ringing the lake where they could watch the activity, de-
scending frequently to harass the rafts of ducks or geese,

flying back and forth above them, making them dive, trying to isolate an individual to make a kill.

A bald eagle has an inability to choose between two or more. As long as the birds kept together they were safe, regardless of how close the eagle came. By harassing them, the eagle tried to separate one from the legions; once its prey was isolated it made its kill. Clumsy as the great bird seemed, it could outfly a little bufflehead and pick it out of the air, if it was a single. On the lake the flocks of waterfowl drew together under harassment, as though they were trying to be cells of a unit too large for an eagle to attack.

It was Nature's way, for the eagles picked up the sick and the inept, rather than the healthy and the alert. Watching the show, I realized for the first time how smoothly the system of predation must have worked for the health of all species before Man came along and dared treat Mother Nature as a fool.

For all I had tried to change Coy into a friend and companion, she was still what Nature had intended her to be, a predator, a veterinarian of the sagebrush and forest around her. Nothing I could do would overcome her instinct to hunt and to kill whatever she could catch.

While the American hunter, the sportsman, is only in the mildest sense selective, shooting the unwary, his tendency is to harvest the biggest and most beautiful, making his selection on a visual basis rather than a genetic one, choosing the phenotype instead of the genotype.

As I sat with Coy looking out over the lake, I thought of all those who persecuted the predator as an animal that competed with us for something we wanted to eat ourselves, ignoring the function of the predator in the system—a function that we humans are unable to perform—and overlooking the benefits.

When we removed the wolf from the scene because it competed with us for what we wanted to eat, we left a vacuum, which the natural system, in its need, filled with the coyote. In nature, if you remove a more specialized animal such as a wolf, it is replaced by a more adaptable one, in this case the coyote. We went from a pack animal with a highly socialized family structure and behavioral patterns, which made it vulnerable to control and eradication, to an animal only loosely organized socially, infinitely adaptable as to food and habitat, and we removed its best functional predator, which was the wolf. Much as we rant and rail, it is a bed of our own making. Create a vacuum and it will be filled by something, but often by something not better than what was there before.

I hugged Coy to me, appreciating her, thinking how lucky I was as a human to have her as a friend. I knew her values on the ranch in devouring rodents and removing the sick and unwary from the population, but there was still something that bothered me. Wildlife biologists concerned with increasing the bag for hunters had cursed the coyote for killing fawns, not weak ones, but big healthy young 'uns, perfectly capable of growing up into magnificent specimens for the harvest.

Granted it takes a long time for genetic changes to affect a population. But what about behavior? I thought back to my uncle's cattle, who hid their calves so that they were hard to find and were perfectly capable of sending a cowboy who got off his horse to rouse the calf right up the nearest tree.

The advent of penicillin and antibiotics changed the cattle industry. Purebred cattle operators started saving calves that otherwise would have died, saved calves that were unhealthy, and more germane to my thoughts on predators, saved calves whose mothers did not possess the instincts to hide or protect their offspring. Some of those calves became herd sires, profoundly affecting the herds into which they were placed.

My cattle are not the mothers my uncle's were. More and more I see animals who leave their calves exposed to wind, rain, and predator. Exceptional is the mother who will put a cowboy up a tree. Unintentionally, I have bred them vulnerable.

And the fawn from the doe who has not hid her off-spring well and has little instinct to protect it, even though it looks big and strong and healthy—do we really want that animal in our game herds?

And if we don't want that animal in our herds, what means do we have to remove it? The hunter who hunts the phenotype, picking out the best? Or the predator who hunts the genotype, selecting prey not on its looks or condition but on its ability to survive?

Except for a few animals left in Minnesota and the West, we have lost the wolf as an aid to making the system work. In Alaska and Canada, the wolf is besieged by those who cannot understand his contribution. He is trapped, shot, poisoned, pursued by snowmobile and helicopter, because sportsmen think there will be more game when the wolf is gone. However, they are only creating a vacuum to be filled by the coyote, and when the coyote is gone, by the invisible predators: disease, starvation, and bad genes.

A cold wind blew down off the mountain, and I hugged Coy more closely, trying to share her fur. Even though the lake was trying its best to freeze over during the cold nights, I had a hankering to be out on her before summer, to be the first man ever since the time of primitive man. From the dark, dusty rafters of a ranch shed, I had taken an Old Town canoe, which my parents had used on their honeymoon in 1911. It seemed fitting that this be the first craft on the lake.

I was only a few feet offshore on my maiden voyage, when Coy discovered that I was leaving her and began yipping and racing up and down the shore. Perhaps she thought I had been captured by some monstrous animal that was now swimming out to sea with what was left of me. Now and then she would scamper for the safety of the timber, but her focus was ever on the canoe.

"It's all right, Coy!" I called.

She cocked her head, uncertain at this cry from the grave. I had to beach the canoe and separate myself completely from its image before she would come near. Coaxing her close, I let her check it out without trying to hurry her. She stretched out her nose gingerly, sniffed it, darted back, tail between her legs, ran two or three half circles, then pretended to ignore it, until a sudden wave moved the craft and gave it unexpected life. She dashed for the woods running full out and desperate.

Within minutes she was back. Lifting her in my arms, I placed her gently on the floor, soothing her with quiet assurances. As I shoved off, she froze, her yellow eyes showing fear, salivating and looking acutely miserable, as though she were carsick.

For a time, she cowed on the bottom, nose on her forepaws, swallowing frequently as saliva puddled on the floor of the canoe. Her head was turned slightly so she could regard my face with one eye, and she was coiled like a spring, ready to jump overboard at the first shadow of my doubt. By the time we had reached mid-lake, however, she had poked her nose over the gunwales and begun to sniff the wind.

I trimmed the canoe, adjusting my weight to hers, and let the canoe drift, using my paddle as a rudder, watching the chill March wind ripple the oatfields of her winter coat. She stood up and put both feet on the gunwale, letting the wind compress the hairs along her muzzle, glancing back at me as though to say "This is fun!"

"Careful, Coy!" I warned.

At that moment, a piece of gray driftwood floated by. Rocking on the waves, it looked like a swimming rodent. The sight was too much for the coyote. She leaned over to snatch it from the water.

Over we went, canoe, coyote, and navigator. I sputtered to the surface. The sudden cold sucked away my breath, and my boots felt like anchors, trying to drag me down.

It was Coy's first attempt at swimming, but instinct told her how. She kept trying to climb my head, as though I were a partially submerged log. Her vestigial dew-claws scratched my face and tore at my ears. Holding her off at arm's length, I righted the canoe in the water. As Coy swam to me, whining in terror, I scooped her up in it. Chest deep in sloshing water, she felt the reassurance of the floor beneath her feet and sat motionless, as though determined to behave. Nausea swept over me from the cold. There was not much time left on my clock. Kicking clumsily, pushing the heavy canoe before me, I began the long fight toward shore.

My muscles numbed, easing the pain the cold brought. I straightened one leg and with it tried to kick the other. The shore looked hopelessly far. For a moment the breeze helped us on our way, then just as quickly reversed itself, so that the squall threw a chop against my cheeks. The waves grew larger, as though sensing our helplessness, frothing with a white hydrophobic foam, tasting me, slapping the canoe away. I had the ridiculous thought that they were *my* waves; I had created this lake, and these were my angry children trying to do me in.

I was beginning not to care any more and giggled with some sort of rapture; there was only the pleasantest sort of fatigue. Maybe since I was dying with a coyote, I would go to the place dead coyotes go, and Coy and I would be together in eternity. I hoped they would be patient with me until I could become a better hunter.

Then, suddenly, my feet felt the soft ooze of the bottom, lost it for a moment, then found it again. The canoe bottom ground hard against a rine of shoreline ice; Coy shot out of the canoe, hit the shore scrambling, and scampered to leave the lake behind.

I retched but nothing came, then peace came over me. I

was back in my childhood; waves lapped pleasantly as they had so long ago in northern Michigan. From somewhere up in the forest, Coy woofed at me and whined as though telling me to hurry. I rose unsteadily to my feet and homed in on the cabin. Once indoors, I stripped off my clothes, left them puddling on the linoleum, and threw myself into bed, pulling the covers tight about me.

But I was wasting precious time. There was no heat in the bed save what my body was able to furnish. With me in my condition of hypothermia, that was precious little; I had already donated what heat I had to the lake.

With one great effort, I threw off the covers and staggered to the stove, turning on the tea kettle.

Two nights later, as I was standing on the shore of the lake thinking about my narrow escape, something bumped me behind one knee. Startled, I jumped and turned just as a large coyote shot back out of reach. It was Benji, and behind him stood Coy, looking very pleased.

I dropped to my knees. "Benji! Come here!"

He eyed me with his usual independence, but as Coy pushed past him to bound into my arms, he seemed jealous and came up to crowd her out of the way. His body felt lean and hard as I hugged him, and his damp, coarse hair smelled heavily of carrion. As I talked to him, I scratched his ears, and he tilted his head to show me where it itched.

Unwilling to let him go until I'd had my visit, I picked him up in my arms and carried him toward the cabin. Uttering a warning growl at being thus profaned, his lips curled in a snarl as he touched my throat with his wet nose.

"Go ahead and bite me," I said quietly, offering a full shot at my jugular. Perhaps the fact that I was not afraid saved me a torn throat. To my relief, he relaxed like a rag puppy and allowed himself to be carried meekly into the cabin.

Once inside, he wriggled from my grasp, his nose

catching every scent in the cabin. Coy had entered with us and now, as though sensing a celebration, trotted to the refrigerator and put both feet up on the door. I fed them each a chunk of cheese. It was only when I lit the Coleman lamp and flooded the room with light that I saw where Benji had lost part of one front paw in a steel trap.

The stub was a red, angry mess. I tried to soak it in a solution of warm boric acid, but the ordeal was too much for both of us. The solution ended up on the bedroom floor with Benji high and dry on the bed. I managed one good shot of sulpha from a spray can, but he snarled at me and meant business.

Coy, however, seemed to think he was bluffing and leaped up beside him on the bed, where he lay licking the medication off the wound. Wriggling up beside him, she kissed him with tiny licks along his jaw. For a moment I feared he was going to deliver up his groceries. He lifted his nose out of reach, then when she edged even closer, he tried firmer means, a noisy, snarling attack that left a puff of coyote fur on the bed and a contrite Coy under the kitchen table. I patted her to keep her occupied. Moments later Benji was asleep, stretched out on my bed as though he had lived there always.

Judging from the condition of his coat, Benji had adjusted well to the wild. There was a piece gone from one ear and some fresh scars across his muzzle, showing he had fought in earnest combat, perhaps against an older male for a female or a carcass. I could not help but wonder what adventures had befallen him during those months, or why he had chosen this particular moment to return.

Sleeping between two coyotes, I spent a restless night. Now and again Benji would raise his head to listen to some far-off coyote howl, then sag down to rest. His wound must have pained him a great deal, for he spent much time dressing it with his tongue. Often he pushed himself up to scratch, thumping up a dust storm on the gray canvas bedcover as he pursued a flea.

I slept, but when next I awoke I was on fire, itching and scratching and doing everything but biting myself. It was morning, and as the sun streamed in my window and puddled on the bed, I saw that the joint was jumping with fleas—a gift from the wild that Benji had brought me. Both coyotes lay quite meekly as I dusted them with malathion powder.

It may be that Benji had brought home his problems to me, both the fleas and the injury. Or it may have been that his foot hurt him too much to hunt, and hunger had forced him back to the cabin, where experience had taught him that living was easy.

Within a week, once the fleas were gone, the foot no longer hurt him, and his belly was full, the old restlessness returned. Early one morning I felt a wet nose at my ear. I fought against waking, but soon the regular click of his toenails on the floor told me that he was pacing with some urgency.

Putting on my slippers, I headed to the door to let him out. There on the porch was a wild coyote. Seeing motion, it scrambled away and vanished like a wraith, leaving Benji making fast laps around the outside of the cabin to pick up the stranger's scent. Once he had found it and sniffed it, however, he relaxed as though in recognition. In front of the cabin, he paused to bark once and howl. Almost immediately, from the forest half a mile away came an answer.

Benji sat and heard the coyote out, then decided to follow. He was in no hurry. Halfway across my newly planted strawberry patch, he dug up a pocket gopher and gave it to Coy, as though a girl couldn't catch her own. Then a newly planted lilac bush caught his eye, and he stopped to chew off a branch.

"No, Benji!" I scolded. It came up louder and harsher than I intended.

For a moment he stood looking at me, his head cocked first to one side, then the other, as though evaluating our relationship and deciding whether it was worth being yelled at. Then he turned his back on my discipline and rules, trotted across the dam, caught a mouse along the bank, ate it, and vanished out of my life forever.

RANCHES ARE DEMANDING MISTRESSES. Although I had managed a winter respite because of the quality and dedication of the two men who worked for me, once calving season had arrived the responsibility was more than they cared to handle without my presence at the business end of the ranch. Coy seemed to understand where we were going, for I had only to open the pick-up door and she was in the seat and ready to be off. As we drove away from the cabin, I saw Don Coyote perched up on the tracks of Big Alice, enjoying a bit of sun. I wouldn't be needing the tractor now for several days, and I was pleased to think that he would have it all to himself.

Coy hit the ranch yard like a March blizzard. She bowled over Mickey, the old cow dog, as he limped up to greet me, landed in the middle of Noah, who was sound asleep in a bed of daffodils, and sent Toby, the porcupine, climbing up to the top of the woodpile, where he sat blinking his one good eye to see what all the commotion was about.

Nose to ground, she rushed about the yard as though trying to pick up scents of her childhood, drove the saucy Steller's jays and camp robbers out of the dog dish, and sent the great blue herons and Canada geese that consider the house lot their domain winging out over the meadows.

With the end of the long winter, I had my own excitements, and with the melting snows came the itch to feel the familiar lurch of old Straight Edge beneath me and to ride the old mare out over the greening meadows. I saddled her and took off.

The geese were shouting their noisy nesting quarrels out on the meadow, and as I rode, I watched five families of greater sandhill cranes sweep over the long ridges of Taylor Butte and end their migratory journey by gliding each to its own nesting territory. I was relieved that the group included a pair of cranes that had nested beside our house spring for fifteen consecutive years.

Spring on the ranch is a busy time for humans as well as for wildlife. The cows were dropping calves like flies, and a scattering of six or eight coyotes stayed in near-constant attendance, dragging off the large pink flags of afterbirth as the cows led their tiny calves away from the dropping area.

"Sonofabitch'll git your calves," the sheepman had told me.

It wasn't happening now, nor had it happened in my uncle's time. We had to be doing something right that other ranchers did not do. Whatever it was, was most likely accidental on our part, probably a combination of several factors and not some obvious thing that I could put my finger on.

Though a bit tattered and ready to shed their long winter coats, our coyotes appeared to be in good condition. I was convinced that, first of all, it paid to keep coyotes well fed. That we did well.

After lunch I set off on Straight Edge, with Coy scampering circles around us, to pay a visit to my worry log. As Coy went running through the woods ahead of me, she surprised a big male coyote patrolling the woods. The startled animal lit out running, though it turned back atop the hill and watched us as though surprised that one of its kind would be on friendly terms with Man.

Tame as Coy was, she was a good specimen for me to observe, for outside of being loyal to me, she went about her life behaving like a wild coyote and was well accepted by them. If she dined occasionally from the dishes of domestic dogs, she preferred wild fare and was perfectly capable of subsisting by her own hunting prowess.

The worry log, though a trifle damp, appeared to have wintered well. I sat in my old seat, enjoying the sun as it sent steam from the log to rise about me and thinking back on my uncle and my children, trout fishing on the river, the wildlife I had known, the lake, and a host of other nice things that had happened to me. Coming up was another year of my management filled with challenges and, I supposed, disappointments, but it beat the hell out of living in town.

Looking down over the meadows, I was surprised at how green the ranch was compared to most. One of the things we did differently than most ranchers was to leave plenty of old feed to protect the new growth in the spring. It was just plain good grass management, but suddenly as I sat on my log, I saw that it was good coyote management as well. The residual forage allowed mice to move farther from their burrows and made them easier to hunt, and even in the snow, the tall grasses made for a stable rodent supply a coyote could count on.

Then too, since our coyotes were cherished, they could feel comfortable hunting on our ranch even in broad daylight. Not only did the high grasses afford them protection, but they also had a sanctuary where they could hunt without being shot at and could concentrate not on staying alive but on taking prey.

Thus, throughout the course of a twenty-four-hour period, our coyotes could spend more time hunting because their diets were not limited to nocturnal species, and their scrounging success was higher than that of coyotes suffering persecution.

The river, too, had to be a factor. The Williamson stayed open during the winter, and there were always areas along the watercourse that were covered with heavy vegetation, and here, whatever the weather, even when heavy snows buried the meadows, every predator, whether he had three legs or four, had a good chance to make a living.

One of the most important aspects of our coyote management, however, was our treatment of dead animals. Every rancher has a certain amount of carrion. Whether you manage cattle well or not, some of them die. From my uncle's day on, we had dragged each dead animal out into the brush and left it for the coyotes to clean up, selecting areas away from calving fields, searching out places where a coyote could feel comfortable. It not only removed the animal from our sight and kept us from getting depressed at its loss, but it kept the banker from knowing just how many we were losing. The idea was to keep the cow alive on paper as long as possible after she was dead.

If we had hated the coyotes, we would have set traps beside the carcasses or laced them with poison, and ended up teaching the survivors to avoid carcasses in the future. Instead, they trusted us. In effect, we trained the predators to utilize carrion and thus brought them through the hungriest periods of winter when rodent populations were lowest.

A hungry predator is a problem predator. By feeding and protecting our coyotes, we were insuring that each

coyote on the ranch lived to a ripe old age. As a resident
of a territory, he would keep strays from coming in, and
by virtue of knowing every mouse run in his fiefdom and
every squirrel town, he could make a comfortable living
without getting into mischief.

There is an old saying that if something works, don't
fix it. It can be applied to predators. If you are halfway
getting along with the coyotes on your place, don't remove one. It will only create a vacuum that will be filled
by an animal you might not be able to tolerate at all. In
most cases, unless an individual coyote gets a hang-up on
killing stock, the existing population is best.

As to reducing livestock losses by removing predators,
landowners are simply trading short-term benefits for
long-range problems. Studies such as H. T. Gier's *Coyotes
in Kansas* (1968) have shown that coyote control puts sur-
viving coyotes into production, causing an increase in the
size of litters and influencing bitches to start whelping at
younger ages. An older bitch can teach a pup to hunt so
that it is less apt to go off into the world unschooled and
hungry, primed for trouble.

As I leaned back against my worry log, I saw overgraz-
ing as one of the most important causes of livestock predation. Short grass makes for poor hunting, unstable
prey populations, and hungry predators. In my travels I
have seen many a ranch; the operators who grazed their
pastures close were the ones who suffered the most serious losses.

I recognized that there was no way I could make
coyote lovers of them. The only thing I could do was to
manage my own land *for* predators instead of *against* them
and hope that someone, someday, would look over my
fences and realize that I had something special going for
me.

With the warming of the weather, Gerdi and the kids
came back from town to help me with the cattle. They
were good riders all of them, and the horses they rode

put Straight Edge to shame. But somehow I couldn't bear to trade the old mare in for a younger model. We went our own pace, the aging wreck and I, with Coy following hard upon her heels. Often Coy trailed so close behind that I went hours without seeing her. Once as I stopped the mare in the middle of a large flat, wondering how on earth Coy had vanished so completely, I leaned over to find her curled up, nose to tail, directly beneath us, in the shade of Straight Edge's belly.

During the week, dawn to dark, I repaired fences, closed gates we'd opened for the winter, and checked irrigation ditches for mouse damage. At the end of the day Coy was still ready for adventure while I could hardly wait for bed.

The coyote put up with me well, but she sensed when Friday had rolled around. Along with Mickey and Noah, she took to sitting and watching the road, waiting patiently for the rumble of the family car as it crossed the cattle guard. Like the others, she was beside herself with joy at seeing everyone. She leaped for kites, chased after bicycles, stole sandwiches and even apples. A sandlot baseball game on the meadow in front of the ranch house brought her from afar. Full of mischief, she stole base after base, dragging them off into the woods. Outfielders had to look sharp or Coy had the ball before them. Her version of the sport was to take the ball into the woods, elude her pursuers, and bury it.

Spare moments, we rushed to the lake and cabin, where Coy's first act was to check the whereabouts of Mrs. Flannigan and keep her honest. Week after week, that lady came out on the shelves with growing impudence and peered down at us, curious to see who would dare to come unbidden into her boudoir.

As April became May, the weather warmed, the grasses grew lush, and we tried out my new system of irrigation. We filled marshes, ditches, and ponds, and let the icy spring waters stand to warm in the sun before we sent them flooding over the meadows. Maybe we still heated the ranch house with a wood furnace, but out on the land I had gone solar, trying to moderate the ranch climate with every bit of solar heat I could store.

Riding down the ranch one day on Straight Edge, I thrilled at the change. Nice things were happening: For the first time I could see a system working on the land. Nature was beginning to achieve some sort of equilibrium. We were at peace with the land, working with Nature instead of attacking her head on.

The ancient marshes were coming back; coarse vegetation sprang up almost overnight, creating a whole new habitat for wildlife—no matter that those grasses did not rate well in the county agent's handbook. Cows got fat on them, and that was what payed the bills.

All day long, the tea-colored marsh waters absorbed the rays of the mountain sun, and all night long they released heat, warming the climate. Where once even sedges and rushes had turned brown with the night frosts, now they grew green and lush, and frogs croaked all night in the warm, moist darkness. For once we had more feed than cattle to eat it.

In those same marshes, a host of birds—yellowthroats; red-winged, yellow-headed, and Brewer's blackbirds; marsh wrens; meadowlarks; rails—moved in to nest and raise their young. We helped them by restoring the marshes; they helped us by taking care of insect problems. The great grasshopper epidemics which had plagued my uncle and me became history. In the traditional grasshopper nesting areas, clouds of birds worked on nymphs, carting them away to feed their young. Hour after hour, day after day, sandhill cranes stood shredding young grasshoppers with their long bills.

We had once relied on poisons; now as we checked the areas where grasshoppers had once turned the land to dust, we found bluebirds, wrens, flickers, sparrow and red-tailed hawks, coyotes, badgers, foxes, cranes, blackbirds, meadowlarks, and even ducks cleaning up the areas. By the time their young were raised, the 'hoppers they'd missed wouldn't have filled a tobacco can.

There is more to maintaining wetlands than backing up water and letting it stand. Certain species of plants are bullies and crowd out competition until there is little diversity left. Marshes tend to choke up with coarse vegetation, which, left ungrazed or unharvested, rots. Periodic drying is essential.

The balance of wildlife species too is important. When we lost our coyotes to poison, the raccoons proliferated until there was hardly a nest on the place they hadn't destroyed. Before the coyotes remultiplied and were again numerous enough to control the raccoons, waterfowl production had been almost nil. Now we had ducks and geese back nesting by the hundreds.

I thought of the other species on the ranch. Without flickers, badgers, trout, deer, or chipmunks, the ranch still would have flourished. But if I took away the coyotes, the whole system fell apart. They were as necessary to the well-being of Yamsi as any tool I owned, including shovel, pick-up truck, mower, hay baler, fencing stretcher, pliers, welding outfit, saddle horse, saddle, rope, medicine, and tractor. In fact, if I were to design a kit for the beginning rancher, a pair of coyotes would have to be included.

There was a difference between my ranch and every other I knew about whose owners complained about coyote damage. Acre for acre, I had three to five coyotes to their one. Yet while they lay awake at night waiting for predators to kill their livestock, I slept like a baby, not hoping, knowing my calves were safe.

The secret, of course, was that I kept my coyotes fed all year round. All year round! Not eleven months. Twelve! When prey was slack, all the carrion they could eat. It wasn't the coyote at all that did the damage. It was the coyote's appetite!

One afternoon in late May, I stood leaning up against a gnarled lodgepole pine, letting the old mare rest and graze a bit, and gazed off over verdant meadows at fat, sleek cattle, so stuffed they didn't need another bite of grass. In forty years I hadn't seen the place look better, and we were running fifty-three percent more cattle than in the years when frost and grasshoppers stole a big chunk out of the yield.

When I had arrived as a boy, there hadn't been that much wildlife; now it abounded in spectacular numbers and diversity. Bald eagles watched for prey along the river, ospreys fished the river's bends, great gray owls hunted the meadows for mice. Not only did Yamsi take care of a host of resident species, but it also made thousands of acres of arid federal forest around it viable for wildlife.

Straight Edge kicked at a fly and interrupted my musings, then dozed off again. As Coy came in off a hunt and plunked herself down beside us, a big coyote followed her, but saw us and left in haste. He looked big enough to be a small wolf.

Coyotes had been looking bigger to me lately. Years of persecution had speeded up the process of selection for the coyote. We had shot off the alfalfa field mouser who was our friend and had left the larger, craftier coyote who happened to be out in the back country hunting larger game, creating a super-coyote more solitary by nature and thus less vulnerable to control. It took over a niche abandoned by the wolf.

Ever since the great mouse epidemic, I had relied less and less on poisoned grains to keep ground squirrels in check, and now I used none at all. The epidemic of these little grass eaters I had feared didn't happen. Populations stayed low and fairly stable, and I came to look on the ground squirrels I had on my land as beneficial, since they fed my varied predators.

The ranch was a great place to see Nature working,

since there were no neighboring land use influences that might affect my animals unless one counted Indians and poachers who nightly spotlighted deer as they came down to water on my land.

Coy put her front feet up on me and leaned against me to be petted. "You guys," I said. "Every time I try to put in a good word about you coyotes to my friends, they get a furtive look in their eyes and their minds spring shut like steel traps. If they talk to me at all, they pour out a list of grievances about how coyotes have done this and that, how you've killed deer, calves, lambs, chickens, pet dogs and cats, birds, and God knows what else. Of course you have. But as long as you make me more money than you cost me, I intend to keep you around. I figure other ranchers could do the same and come out winners if they'd just learn how to coexist."

Coy tipped her head as she looked up at me seemingly puzzled at what I was mumbling about. Unless she was hunting prey, her attention span was short. She forgot me entirely when a pine squirrel rattled the bark above our heads; apparently the sight of the squirrel started her thinking about food. As I mounted Straight Edge and turned her head toward the barn, Coy went trotting down to the ground squirrel town below us to look for some action.

THE SUMMER THAT FOLLOWED was the driest in recent history.

"You caused this whole disaster, Dad," Dayton grinned. "If you hadn't closed the dam to fill the lake, the weather patterns would be normal. The neighbors and town folks are all laughing at us, calling our lake 'Maybe Lake,' telling each other it's never going to fill."

"Look at the bright side," I replied. "If we hadn't built that dam, by now there wouldn't be enough for a chickadee to drink, yet we've got three-quarters of a lake and that's lots better than what most folks in this part of the country are faced with."

All over the West, water holes were drying up, cattle were wearing their hooves to a nub hunting feed, and here the neighbors were laughing because our lake hadn't filled to the brim.

Having fought my way through a series of small disas-

ters lately, I considered the drought to be overly cruel. I
needed the satisfaction of having something work out as
planned. If I squinted, it made the lake look larger and
more significant, but I couldn't go around squinting all
summer. Like the birds and animals around me, I had to
take what Nature provided and be content.

It took me a few days of seeing other people suffer to
realize just how lucky we were. A dry year is also a cold
year. But the effect of the lake both upon the local climate and the wildlife populations was startling. During
the spring, that body of cold water had kept the plantlife
from coming out too early. Now the lake was acting as a
reservoir not only of water but of solar heat, guarding
each lakeside plant with gentle fingers of warm, moist air
from nighttime frosts.

A new lake is chock full of nutrients, chemicals in solution as yet unspoken for, which boom populations of all
manner of organisms, simple and complex, from tiny
microorganisms to three-inch dragonflies and monstrous
pond beetles larger than a man's big toe. Out of nowhere
came mayfly hatches staggering to behold, as though the
eggs had lain there since prehistoric times just waiting for
the right conditions to let them hatch.

Once that brew of lake water began to work, it was as
though an invitation had gone out to wildlife for fifty
miles around. The only lonely water for miles, it attracted everything from hummingbirds to bears. At times
there was hardly a foot of surface not patrolled by a flitting swallow. In the evenings tree swallows descended in
clouds, covering the scattered dead trees near the cabin
until branches broke under their weight and the roost lost
its visual identity. Long after dark their chittering kept
me awake.

Wilson's snipe rose high above the evening lake to
winnow to their wives, wing singing as they dove
through soft summer air. Nighthawks boomed, hawking
insects for their young scattered round and about the

floor of the forest. Mourning doves raced each other in from off a million arid acres to drink. Canada geese sailed their growing families through molt, making my mornings raucous with their clamor. At each outburst of geese, coyotes voiced their own hysteria from the surrounding wooded mountains.

Under the eaves and between the walls of the cabin nested a host of birds in near harmony, flitting in and out of knotholes in the weathered clapboard: mountain bluebirds; red-breasted, white breasted, and pigmy nuthatches; mountain chickadees; brown creepers; flickers; and tree swallows lived in the walls. Barn swallows selected ledges above window and doorjambs, while cliff swallows plastered the eaves with mud gourds. A family of bats moved into the stone chimney, buzzing and squeaking endlessly through the night. At dawn a new chorus arose from the wall a few inches from my pillow as a hatch of starlings clamored for food.

Perhaps Coy mistook the uproar for a new bit of audacity on the part of Mrs. Flannigan. She leaped from my bed and tried to isolate each new chorus of squeaks in the wall. Each time a parent bird arrived with a beak full of insects from along the lake, Coy worked her way closer and closer to the clamor, until at last she leaped past my pillow and began to tear out the wall with paws and teeth. She had already showered the bed with splinters when I banished her from the room.

The birds had invaded Mrs. Flannigan's inner sanctum between the walls, and that drove the lady indoors for good. She lived on the shelves out of Coy's reach, and whenever I rustled cellophane I could see her whiskers working from behind a cannister or jar as she sniffed the air for a meal.

I was less successful in preventing Coy's war against the bats, which ran a hotel for the transient hordes flitting nightly across the lake. Coy's attacks on the kitchen wall not only sent the bats into a chorus of buzzes, but brought a sifting of bat droppings down through cracks in the wall. Dutifully, each morning I swept them up and composted them for eventual use in my garden, a special gift from Nature not available at my local garden supply store.

As if Heaven were not close enough, a pair of rare black-backed three-toed woodpeckers built a nest just outside my cabin door. Not quite believing my good fortune, I watched them make their first exploratory tappings on the bark of a lodgepole pine and go on to excavate a nest, incubate eggs, and raise three young.

Each morning I took my binoculars to a rocky point overlooking the lake to check for new arrivals. At seven o'clock one morning, a pair of ospreys arrived to check the lake for fish that had yet to be planted. They were persistent; it was as though they were thinking to themselves, "In a lake like this, there just *have* to be fish." Now and then they powered from the heights, talons lowered, and broke the mirrored calm. But after floating awkwardly for a few moments, half swimming, half enjoying a bath, they flapped up dripping into the air.

As the ospreys flew off over the ridges toward the Sprague, where the fishing was better, the bald eagles sitting on their lookout trees took over the lake, making forays over the ducks and geese to search among the thousands for any who might be ill or unwary. High on a lodgepole behind the cabin, a red-tailed hawk raised its screaming brood on ground squirrels, while in a hollow tree west of the dam, a kestrel sat patiently in the sunshine, guarding a certain secret hollow in the snag below.

For weeks a pair of curlews came to a sandbar to feed. I had hopes of their nesting there until one morning, as I happened to be watching, a peregrine falcon came flinging out of nowhere to dissolve one of the feeding birds in a puff of feathers. Any morning, I might watch a gentle great gray owl hunting chipmunks or glimpse a goshawk flying swift and low to baffle the distant early warning systems of some hapless rodent caught away from its burrow, insuring genetically that future generations of squirrels would be more wary.

In the marshes forming at the shallow end of the lake, a pair of sandhill cranes fed quietly and unobtrusively for several days. Then one morning I noticed that one had disappeared. Searching the area, I found the hen crouched low, head stretched forward, on a nest of reeds and grasses, a bulky bird-made island some thirty feet offshore.

One day at dawn a strange wail sent Coy sailing off the bed with a loud "WUFF!" Again I heard it, and Coy answered with another bark and a howl.

With Coy leading, I dashed out the door of the cabin toward the lake. Again the eerie call drifted across the dark waters. I saw it then, swimming half submerged a hundred yards offshore. A loon! The first ever on my lake! The first I'd seen since my boyhood in northern Michigan.

Looking for fish, it swam from one end of the lake to the other but found my hospitality lacking, and it soon beat heavily across the water, circled through the air until it had gained height, then passed over the ridges heading north toward some ancestral ground.

When you've had a loon and lost it, a lake can be a lonely place. I resolved to plant fish so that, when it returned again some year, I might be a better host.

As summer waned and autumn came, great flights of ducks and geese found both the new marshes on Yamsi and at the lake a haven. The desperate drought, which had dried up so much of the West, had only intensified their need and brought them miles off their regular flyway patterns to rest with us. One morning at the lake, when over five thousand ducks, geese, and swans rose from the lake to escape a pair of harassing bald eagles, I thought back on my uncle's tales of the early days when waterfowl darkened the skies. The whistling hordes swept over me like a giant cloud.

Often as I rode old Straight Edge down along the Yamsi meadows, I marveled at the peace and tranquility there, enjoying the fat beef animals grazing on meadows still lush with feed even though summer was gone. The warm waters of the marshes were yielding only reluctantly to the cool night breezes settling off the mountains, and still kept the killing frosts at bay.

This year, instead of a grasshopper epidemic there were only a scattered few. The birds had done their work well. Predators had held the ground squirrels and meadow mice to levels we could work with. Nature, which was holding much of the rest of the country in thrall, seemed at last to be on our side.

Straight Edge was getting old. The sons and daughters she had produced through the years formed the bulk of our remuda now, but she seemed still to enjoy easy rides up and down the valley, shying at the same stumps and imaginary bears she had worried and fretted over for years. I spoiled her badly now, letting her stop to graze on special clumps of grass I knew she'd enjoy. Soon it would be time to turn her out forever and let her graze out her remaining time. She'd seen me through some adventures and had been as patient with me as now I had to be patient with her.

As I rode, I saw Don Coyote often. His attachment for the old tractor was seemingly undiminished, and I had no doubt but that he spent the heat of the day and the fury of the storm beneath it. Now and then Coy would play coquette with him, but when she became too familiar, the old grouch drove her away. Now and then he followed us, but at a distance, using my presence to explore other coyote territories than his own, as though hoping that in my company the watchful coyotes would behave and grant him right of passage.

One day I sat astride Straight Edge, one of my knees hooked over the saddle horn, watching lazily as Coy hunted the tall bleached grasses. The V of the Don's ears were all I could see of him as he spied on us from a knoll overlooking the lake. I looked around me with ever increasing wonder at the harmony about the place. Despite the season's grazing, there were still places where grass touched my stirrups, as it had back in the days when my uncle had first seen this lovely land.

Out on the meadows, Coy made a wild pounce and came up with a fat pocket gopher, and the Don broke his cover to come over to see what she had killed. The tall grasses screened his deformities, and he looked like any other fine specimen of coyote male. I was struck by how much landowners were missing when they failed to make peace with their predators and the land.

I FELT GUILTY DRIVING Big Alice through a bit of our own private forest some distance from the lake. Those lonely glades were almost healed now from the early ravages of Man. The area was trackless save for one small, meandering forest road even now beginning to choke up with young pines.

A few months before, while running with Coy, I had found, ringed with aspens, a small seep spring which trickled its short life down a draw and was lost forever in the pumice sands. A few passes with Big Alice would be all that was needed to recreate a small pond, which had silted in, where wood ducks might nest and wild things slip in unexposed to drink. My final act would be to close the road through the woods permanently, thus setting up a block of forest where wildlife would have privacy.

Big Alice had troubles deep within her womb, but she tried. As we sputtered and clanked through the woods, I

struggled to remember how she had sounded years ago—
instead of a death rattle, the purr of a contented cougar.
Add up all that dirt she'd moved with that once-shiny
blade, and it would come to quite a mountain.

The salesman who had come out from town to examine her had only looked embarrassed when I mentioned a
trade. The time to trade her in for a newer model had
passed.

"Cost more to haul her to town than she's worth," he
said. "Best drive her till she falls apart, then use her as a
corner brace for a fence."

The end came sooner than I expected. I'd finished
cleaning out the pond below the spring, blocked the
road, and huffed and puffed up a ridge, hoping to slip
back out of the forest through an open stand of pines so
as to leave as little sign as possible. Suddenly, the radiator
sent an eruption of steam ghosting a hundred feet in the
air. It took me an hour to cool her down again. Time and
again, I packed water in my hat and climbed the long
ridge from the spring, hoping each trip would be the last
I needed to cool her fever.

We set out again bravely enough, but as we struggled
across a pretty little knoll ringed with trees, I paused to
look down on the lake before me and to reflect on its
beauty.

Perhaps Big Alice liked the scene well enough herself,
for she chose that moment to die. A great shudder seized
her; with a huge bang, she threw a connecting rod right
through the side of the block, and her hot, black lifeblood
gushed out into the snow. A blue donut of exhaust, her
last ever, sailed higher and higher in the pines until a
thermal moving up the west shoulder of Calimus Butte
wafted the wraith away. I watched open-mouthed as
though I were watching her soul depart.

Taking a small target pistol from my belt, I finalized
her retirement from hard labor by shooting her through
the radiator. A week later, when I rode by Big Alice to

remove some wrenches from her toolbox, I saw fresh tracks in the snow where Don Coyote had been following her spoor and had limped down off the ridges to claim her. Peering beneath her hulk, I could see sand on snow, where he had dug himself a new den.

Coy was off hunting when I left the cabin to go up to Yamsi for Christmas Eve. That night a big dump of snow made roads impassable, and for nearly a week my family and I were snowed in together at the ranch. I hoped that Coy would find prey and take care of herself without me.

My daughter Ginny peered out the front windows at the storm raging outside. It was a great whiteout; not even the yard fence was visible. "We've got you now, Dad," Ginny teased. "No more sharing you with Coy until the weather clears and the plows open the roads."

I paced the floor as nervously as Coy when she wanted to be let out. It had been quite a day of being shut in. Already Gerdi had swept under my feet a dozen times, and three times I'd been caught with both elbows on the table.

"Wouldn't it be nice, Dad," Ginn remarked as I tried to bury my nose in a book. "Wouldn't it be super if Coy hit it off with a wild coyote—like Don Coyote, for instance! Maybe that old three-legged Casanova is lonesome. I haven't seen his female friend around since deer season."

"Don Coyote! No way. He's much too old for her, and missing a leg and a tail. What sort of a life would they have together with her always having to stop and wait for the old man to catch up?"

"Dad!"

"And what do we know about his background? I mean *really*. What kind of coyotes were his parents? Gypsies, maybe, with no sense of home range."

"Dad! You're jealous, that's what! You don't really want to give her up to the wild, do you?"

"Got to be so careful these days," I said. "For instance,

there's that male high up on the shoulder of Calimus Butte. Not a bad specimen, but the country he runs in is too vulnerable. When I was a boy it was wild country. Now the Forest Service has cut the country to ribbons. Build too many roads—superhighways through the woods—those guys, and with no reason. Destroy wildlife habitats, bring in road hunters and spotlighters, and take good land out of timber production forever. And those clear-cuts they did on steep pumice slopes right along the Williamson—!"

"Dad! Back on the subject. We were talking about Coy, remember?"

"And that family that lives south of the lake toward Long Prairie. A structured family almost like wolves. Coy would just be a handmaiden for the cranky old bitch that runs them."

I glanced out the window hopefully, wishing that the storm had eased so I could go out to check the horses and escape a conversation painful to me. But the storm bore down harder and harder.

"Father!" Ginny snapped. "You can't escape. It's time you faced the facts. You've tried your best to domesticate Coy, to make a tame animal out of a wild one. She's not a dog, you know. She's a wild animal and would be much happier out in the woods hunting for a living. Of all people, I would expect you to know that!"

For a long moment I stared at her, not knowing quite how to defend my honor. She was right, and I had been a fool not to see it. I had meddled in Don Coyote's life, and now I was meddling in Coy's. I gave Ginny a hug in lieu of words, and as I looked over her shoulder at Calimus Butte, the clouds lifted and exposed the ridges sparkling under a cloak of snow.

When the snowplow finally opened the road, I was the first one to use it. I parked my pick-up as close to the cabin as I could get and snowshoed in, pausing every half mile to puff and blow. The snow west of the dam was trackless, but as I crossed the dam, I saw Coy sitting on the front porch amidst the ruin of several snowshoe hares and pine squirrels. There was no sign of Don Coyote, but the tracks indicated that Coy had slept under the porch, camping there for several days as though waiting for me to come home.

Her coat was a mass of ice balls, which clicked and rattled like castanets as she romped about in her joy at seeing me. Plunging through fresh new drifts, she immersed her muzzle until only her ears and back were visible. I felt somehow reassured. Coy might stay on with me of her own volition. No coyote male in his right mind would fall in love with that ice maiden!

I melted snow water for tea, and Coy lay on the floor before the stove waiting for the ice balls to melt. Once she had groomed herself, however, a restlessness seized her, and she got to her feet, leaving a small spot of blood on the floor, the badge of her first heat. When I went out for an arm load of wood she bulled on past me, shoving me aside, and vanished into the gloom.

The night seemed endless. Now and then, from the crawl space beneath the cabin, I heard barks, growls, and whines as she guarded the entrance against a visitor. Standing on the porch in my nightshirt, I tried calling her in, but all I accomplished was to set the wild coyotes to howling, and before the last howl died on the ridges, I was half frozen and ready for the warmth of bed.

During the night, another snow fell, painting out the stories daylight could have told. I half expected Coy to be sitting at the door waiting to be let in, but the snow was unbroken by tracks.

That afternoon I saw Coy and Don Coyote together. They were sniffing out mice along the far shore of the lake, and my binoculars brought them up so close I felt guilty of spying. Coy cornered a muskrat under thin ice and killed it. Dragging it to high ground, she stood on the prize with both feet and tore at it with her teeth, while the Don sniffed the quarry from a few feet away but made no attempt to take it from her.

Once the muskrat had been devoured, Coy backed away, allowing the Don to gobble up what few scraps remained. When he had finished, she tried to play with him, but, fangs bared, he snapped at her, and I thought that it was unlikely that there would ever be anything between them.

Now, however, she fawned before him, wriggling up to him, licking his face, his ears, his neck. He held his nose up away from her, as though he considered all this display of affection a bother.

As she lay on her back in submission, with one hind leg cocked, he sniffed her tail, and his stub wagged stiffly just once or twice to show his interest. No great outpouring of emotion. She tried to rise, but he drove her back down into the snow and stood over her.

Stiff-legged, he limped away, pausing to urinate on a dead limb protruding from the snow. He did it so easily—a Cary Grant of Coyotes. Unencumbered by a leg, he made it work to his advantage, lifting his stub smoothly, carelessly, high in the air. Reading back his message, he appeared satisfied, and with Coy following at a respectful distance, moved off toward Big Alice.

It was midnight when she scratched to get in, then lay beside my bed on her rug, dreaming coyote dreams as she slept. In the darkness before dawn, I was startled by a growl outside my cabin. Roused from her slumber, Coy rushed to the door while I staggered up and lit a Coleman lamp. In the puddle of yellow light that spilled out onto the snow, I caught a glimpse of the Don, but he sprinted for the shadows.

Coy jumped into the snow ready to play, but the Don darted from the shadows of a pine, hit her with his broad chest, and pawed her down into a drift. Subdued, she lay quietly, lips in a mock snarl, tail thumping great slow beats as he checked out her condition. Suddenly she squirmed up from the snow and began to run, sprinting figure eights around him. As the Don tried to whirl to keep up with her, she leaped over his back. No match for Coy, the Don took shortcuts across her circles and managed to bowl her over again. But she filled his grinning face with flying feet and escaped.

Feeling that I had no right to watch their courtship, I went back into the cabin and put another piece of lodgepole in the stove, then sat sulking and feeling totally deserted. Some time later she scratched at the door again to be let in.

Neither of us slept. Coy seemed driven by devils. All night long, her toenails clacked across the floor as she paced restlessly back and forth, caught up by the silliness of her heat. In that small cabin that night she must have traveled well over a hundred miles.

Outside in the darkness, coyotes yapped incessantly. I stood out in the snow in my nightshirt and growled my fiercest growls to frighten them away, but it did no good. The quarreling persisted. Shining my lantern out again to illuminate their world, I saw not one male coyote but four. The shaft of light instead of frightening them seemed to throw them into pitched battle while Coy stood off to one side. She looked pleased by the disturbance she had caused.

In the morning the battlefield looked abandoned, but as I set my hand to the door latch, Don Coyote rose suddenly from behind a log where he had been lying in wait. Coy saw him from the window, barked, and raced back and forth in a frenzy.

"Hey," I said. "Settle down! I'll let you out in my own good time. Let that old lover out there cool his heels in the snow until I'm good and ready to open the door."

I ignored her warning growl and tried to reinstate my hold over her by giving her a hug. So swiftly I hardly saw her move, she slashed out at me, and buried her fangs in my upper arm. As I screamed out in pain and surprise, she whirled and sailed through the glass on the door, pushing it out with her chest.

The two lovers ran off together. Soon, as I was bathing my wounds and fighting the shock that follows a biting, I heard them howl togther from the ridge near Big Alice—bloodcurdling and primordial. The full spectrum of coyote experience was reflected there, sometimes joyous and sometimes sad: puppyhood, social games, a convocation of friends or lovers, the mournful sound of lonely times, the excitement of estrus, love, pairing, motherhood, and the hunt.

I did not see Coy for a week. And then one day as I was trying to cut some wood, cursing having to work with my arm in a sling, I looked up to see her sitting quietly in the snow watching me.

I stood there looking, wanting to call out, but too proud. For a moment, I went back to my work, punishing her, and when I looked again she was gone.

"Go then, Coy," I whispered. "Take off with that old grouch if you want. With him to teach you, you'll make it in the wild just fine. Scram! Get out of my life. Go to whatever destiny awaits you!"

Then I turned and walked away. If she watched me go, I never saw her. But I hoped she sat somewhere on the ridge, looking down on me and feeling sad. Already I missed her, but I kept on toward the lake, and soon her spell seemed to lift from me. I was walking forward into a new awareness, a higher consciousness than before.

Everywhere I looked I saw Nature working, and I perceived everything from a new perspective: the methane rising from the bottom muds of the lake to pool under the ice, the water strider patiently dormant but alive in the clear glass at my feet, the Townsend's solitaire's silvery music, the pine seed in the clearing, gathering heat from the weak sun to melt itself down for a head start on spring in the fertile mineral soils.

From the ridges I heard a coyote howl, reminding

other wild creatures that the doctor was about, and for the weak, pain and suffering would not go on much longer.

Proudly, with my children, I had built a lake, but we had done nothing the beaver had not been doing for eons. In living with a coyote I had only come to understand the limitations of my species.

Great flakes of snow melted like teardrops on my face as I trudged home to wait out the storm in the shelter of my cabin. Once, just once, I thought I heard coyote music from behind me on the ridges, but it was soon muffled by the approaching storm. By morning, all tracks would be erased forever. Given a fresh start, the gray polluted snow-forms of today would become larger and immaculate, defects lost, resculpted to other artistic creations.

Lying in bed that night I listened for a time to the wind playing along the eaves of the cabin. I felt an urge to howl one last great mournful coyote dirge, but I choked it back. Tomorrow, whatever the storm brought, I would snowshoe out to see my family. Mrs. Flannigan squeaked as she pattered unafraid from behind the wood box and munched some crackers I had left for her. In the stove, a log collapsed into embers, and I slept.

About the Author

When Dayton O. Hyde's book *Don Coyote* was first published in 1986, it was received as a groundbreaking wildlife story and eventually honored by the American Library Association as one of the ten best books of the '80s. For decades, ranchers and sheepmen, as well as the U.S. Government in its "pest eradication" programs, waged an unfair war on coyotes. Hyde put the lie to the myth of the coyote's predation on livestock, and he did so convincingly with his very personal story of befriending coyotes on his 6,000-acre cattle ranch in Oregon's Klamath Basin. In this winning account, he credited the coyote with controlling the rodent population on his ranch (for the first time ever), with the grass growing higher than ever, and with his cattle remaining untouched.

Hyde has been called a rancher's rancher and a naturalist's naturalist. He holds firsthand knowledge of the hard work and hardships of ranching, together with having a dedication to the principles of conservation and sound ecology. In addition, Dayton Hyde is a photographer and essayist, and the author of fifteen books, including *Sandy, Yamsi, Wilderness Ranch,* and *The Major, The Poacher, and the Wonderful One Trout River.* For his unique contributions to conservation and to enhancing our understanding of the wild, he has received numerous honors. A few of his more recent awards are the Honor Award from the Whooping Crane Conservation Association, the Environmentalist of the Year from the National Cattlemen's Association (Region 7), the Dutton Literary Award, The First Hero of the Earth award from Eddie Bauer, Inc., and the Lavin Cup Award for Equine Welfare from the American Association of Equine Practitioners.

Hyde's seasoned advocacy of the environment has led to numerous television appearances on "Good Morning, America," "20/20," the Discovery Channel, a PBS special entitled "Wild Horses: An American Romance," "National Geographic Today," and many more. His stewardship of the land and his ongoing curiosity about the creatures that inhabit the land he has nurtured have been guiding principles in Dayton Hyde's life. After ranching in Oregon from 1950 to 1988, Hyde moved to South Dakota. There, he became the founder and president of the Institute of Range & The American Mustang. Under the auspices of this organization, he now runs the 11,000-acre Black Hills Wild Horse Sanctuary for the protection of wild horses.